# **About** Island Press

Since 1984, the nonprofit Island Press has been stimulating, shaping, and communicating the ideas that are essential for solving environmental problems worldwide. With more than 800 titles in print and some 40 new releases each year, we are the nation's leading publisher on environmental issues. We identify innovative thinkers and emerging trends in the environmental field. We work with world-renowned experts and authors to develop cross-disciplinary solutions to environmental challenges.

Island Press designs and implements coordinated book publication campaigns in order to communicate our critical messages in print, in person, and online using the latest technologies, programs, and the media. Our goal: to reach targeted audiences—scientists, policymakers, environmental advocates, the media, and concerned citizens—who can and will take action to protect the plants and animals that enrich our world, the ecosystems we need to survive, the water we drink, and the air we breathe.

Island Press gratefully acknowledges the support of its work by the Agua Fund, Inc., The Margaret A. Cargill Foundation, Betsy and Jesse Fink Foundation, The William and Flora Hewlett Foundation, The Kresge Foundation, The Forrest and Frances Lattner Foundation, The Andrew W. Mellon Foundation, The Curtis and Edith Munson Foundation, The Overbrook Foundation, The David and Lucile Packard Foundation, The Summit Foundation, Trust for Architectural Easements, The Winslow Foundation, and other generous donors.

The opinions expressed in this book are those of the author(s) and do not necessarily reflect the views of our donors.

# BIRD MIGRATION AND GLOBAL CHANGE

# Bird Migration and Global Change

George W. Cox

**ISLAND**PRESS

*Washington | Covelo | London*

ISLAND PRESS is a trademark of the Center for Resource Economics.

Design and typesetting by Karen Wenk
Printed using Galliard


Library of Congress Cataloging-in-Publication Data

Cox, George W., 1935–
Bird migration and global change / George W. Cox. — 1st ed.
p. cm.
Includes bibliographical references and index.
ISBN-13: 978-1-59726-687-1 (cloth : alk. paper)
ISBN-10: 1-59726-687-6 (cloth : alk. paper)
ISBN-13: 978-1-59726-688-8 (pbk. : alk. paper)
ISBN-10: 1-59726-688-4 (pbk. : alk. paper) 1. Migratory birds—Adaptation.
2. Migratory birds—Effect of habitat modification on. 3. Climatic changes. 4.
Global warming. I. Title.
QL698.9.C69 2010
598.156′8—dc22

2010004093


Printed on recycled, acid-free paper ✺

Manufactured in the United States of America
10 9 8 7 6 5 4 3 2 1


Keywords: bird, migration, global change, climate change, global warming, adaptation,
evolutionary change

# CONTENTS

Will the great migrations of birds survive? Many biologists believe that migratory birds are at greater average risk of extinction due to changing climate than are resident species. This idea is based on the fact that migratory birds depend on different habitats and resources in different locations during their annual cycle and on the fact that the failure of any of these could be fatal. A second general belief is that the climate is now changing at a rate faster than birds have experienced in the past and that evolutionary adjustments by birds may not be able to keep pace with this change.

The thesis of this book is that many migratory birds exhibit a high degree of ecological and evolutionary adaptability and that many are now showing rapid adjustment to climatic changes. A subsidiary thesis, however, is that for other species, climate change can severely constrain or prevent full ecological and evolutionary adjustment, putting their survival at risk. The objective of this book is therefore to evaluate the capacity of migratory birds to respond to the challenges of changing climate.

The book is a review, synthesis, and interpretation of recent scientific literature on migratory birds and their responses to changing climate. We begin by characterizing the patterns of migration shown by birds of different ecological and taxonomic groups, and the frequency of migration in different world environments, from polar regions to the tropics. The nature of climate change throughout the world is examined, and the impacts of this change on bird habitats are evaluated. The special threats of change to particular habitat types and resources utilized by migratory birds are examined and the constraints on responses by birds considered. The particular patterns of influence of climate change on migratory land birds are considered for different biogeographic regions and for special bird groups, including raptors, freshwater birds, and marine birds. Finally, the capacity for ecological and evolutionary adjustment is examined, and the adequacy of current conservation programs is evaluated. No treatment of bird migration has pulled together these diverse topics, which are of critical importance to human efforts to protect biodiversity.

Most of the topics in the book have been addressed for specific groups of birds or for birds in specific geographical areas. Some topics, such as the impacts of climate change on migratory tropical birds (Chapter 10) and evolutionary adaptability of water birds (Chapter 18), have received only spotty attention other than that directed at a few species. Other topics for which no recent syntheses have been attempted include worldwide alteration of migratory bird habitats (Chapter 3), migratory land birds of the temperate Southern Hemisphere (Chapter 9), shorebirds (Chapter 12), waterfowl (Chapter 13), oceanic birds of the North Pacific and Southern Hemisphere (Chapters 15 and 16), evolutionary ability of land birds (Chapter 17), and overall capacity for ecological and evolutionary adjustment (Chapter 19). Recent symposia have tended to focus on land bird responses to changing climate in North America and Europe, but interest is increasing rapidly in Australia. We really know little about how migratory land birds of South America, eastern Asia, and Africa are responding to climatic change, but all indications are that major changes are occurring.

Birding has become not only a hugely popular outdoor activity for millions of people but also an activity that brings people face to face with biodiversity and the threats to its survival. I have not met a birder who is not concerned about protection of bird habitats and about the impacts of changing climates on birds, whether resident or migratory. I hope that this account increases awareness of the changes that migratory birds are experiencing and stimulates readers to participate in efforts to keep the great migrations in motion. See the appendix, as well as www.birdmigration.org, for a list of the common and scientific names of species discussed in the text.

I am indebted to many friends and colleagues who have reviewed portions of the book, especially Ben Becker, Keith Bildstein, Christian Both, Richard Brewer, Lynda Chambers, Glenn Conroy, Hugh Dingle, George Divoky, Peter Dunn, Fred Gehlbach, Frank Gill, Phil Hockey, Steve Oberbauer, Jane Phillips-Conroy, Robert Ricklefs, Kurt Riitters, Tim Sparks, Nils Stenseth, Nils Warnock, and S. Joseph Wright. Many others graciously took time to help me find literature for particular regions or topics: Tatsuya Amano, Ken Chan, Tim Coppack, Victor Cueto, Tom Ebert, Ken Green, Louis Hansen, Stuart Hurlbert, Alex Jahn, Leo Joseph, Brian Linkhart, J. Alan Pounds, Michael Scott, Kim Scribner, Rob Simmons, Liliana Spescha, Wayne Trivelpiece, Rick Wessels, Ed Willis, and Ben Zuckerberg. I especially thank Darla Cox, my wife, for sharing many birding experiences and for reviewing many of the chapters of this book.

George W. Cox
Santa Fe, New Mexico

# PART I

# Introduction

Global climate change is now affecting all major groups of organisms and changing the dynamics of ecosystems from the tropics to polar regions. Bird migration, one of the most complex and intriguing patterns of adaptation to climates that show seasonal changes, is certain to be affected by global change. In this section we shall examine worldwide patterns of bird migration and identify the potential patterns of response by migratory birds to this challenge.

# Chapter 1

# *Bird Migration and Global Change:*
# *The Birds and the Issues*

Birds are the most mobile of vertebrates. Whether they fly, swim, or run, their ability to cover great distances quickly enables many species to exploit different areas at different seasons in the annual cycle and at different stages in their life history. This is most evident in those that fly, but flightless penguins roam the southern oceans far from the islands and coasts that support their breeding colonies. Even a few flightless land birds have well-developed migrations. Emus[1] in arid Western Australia move hundreds of miles in response to rainfall and vegetation conditions. In the forests of New Guinea, Dwarf Cassowaries track seasonal fruit production with altitudinal movements of a thousand or more meters.

Changing climates have special implications for migratory species. Migratory birds depend on specific habitats and resources in different geographical areas at different phases of their annual cycle. How are these complex patterns of habitat and resource use being affected by global climatic change? Is the ecological and evolutionary adaptability of migratory birds adequate to keep pace with changing climates and landscapes? Are future changes in climate likely to cause extinction of many of the world's most remarkable species? These are basic questions that we shall address.

---

[1]Scientific names for species mentioned in the text are given in Appendix 1.

## Migratory Birds: Ecological Patterns

Migratory birds occupy all of the world's major environments, and their movements are extraordinarily diverse. Land, freshwater, and marine birds show well-developed migrations, as do birds of climatic zones ranging from polar regions to the tropics.

For land and freshwater birds, migrants vary enormously in the distance they travel, in the regularity of their schedules from year to year, and in the fraction of individuals that are migratory in different parts of their ranges. Short-distance migrants range from those that show local habitat shifts or altitudinal movements within a small geographical area to those that make intracontinental movements of up to a few hundred kilometers. Where I live in northern New Mexico, for example, the winter Dark-eyed Juncos at our feeder are a mix of latitudinal migrants from the Pacific Northwest and northern Rocky Mountains and gray-headed birds that are altitudinal migrants from nearby mountains. Weather conditions of the particular year influence many of these short-distance patterns. Frequently, only a portion of the population of a species in a given locality shows short-distance movements, a pattern termed partial migration. Over the geographical range of many species, local populations range from being completely resident to partially or fully migratory.

Birds of many different taxa and species at all latitudes exhibit short-distance movements. Birds of mountainous regions, from high latitudes to the tropics, show altitudinal shifts, with many such movements by tropical species yet to be documented. Many birds of semiarid regions in Africa, Asia, and Australia, characterized as "nomadic," show opportunistic movement patterns that enable them to utilize areas that have recently experienced favorable weather conditions.

Long-distance migrations of land and freshwater birds involve flights across major climatic zones, and often between continents or hemispheres. These movements are often more regular in timing than those of short-distance migrants. Strong-flying raptors and shorebirds perform some of the longest migrations. Radio tracking has shown, for example, that Bar-tailed Godwits fly nonstop for distances of 10,000 kilometers or more between breeding areas in Alaska and wintering areas in New Zealand and Australia. The Swainson's Hawks that occasionally fly over my home in New Mexico in spring spend the winter on the Argentinean pampas. Many small songbirds, however, make intercontinental migratory flights of thousands of kilometers.

Seabirds show diverse patterns of migration, as well. Some are short-distance migrants that disperse from coastal or insular breeding colonies to neighboring oceanic regions up to only a few hundred kilometers distant. Others, such as the Arctic Tern, are long-distance migrants that fly thousands of kilometers from breeding areas to nonbreeding ranges in distant oceans, sometimes in the opposite hemisphere.

## Migratory Birds: Taxonomic Patterns

About 9930 species of birds exist worldwide, belonging to 204 families. From many sources in the literature, I have compiled a preliminary list of migratory birds, a task not as easy as it might at first seem. Many species that are commonly regarded as permanent residents are really partial migrants in some parts of their ranges. In northern New Mexico, for example, the Steller's Jay, considered by most reference books to be a permanent resident, is a partial migrant, with many birds moving to lower elevations in winter. Our understanding of the seasonal movements of tropical species, especially altitudinal movements of those of mountainous regions, is still very rudimentary. My survey of migratory species indicates that at least 2600 bird species of 141 families show some type of seasonal migration or substantial nomadism (Table 1.1). This corresponds to about 26.2 percent of all bird species, a figure that is sure to increase as we learn more about bird movements in regions such as eastern Asia, much of Africa, and mountain areas throughout the world. My estimate also substantially exceeds the estimate of 19 percent of migratory species presented by BirdLife International in 2008.

The frequency of migration varies widely among different groups of birds. Members of a few families of tropical birds, such as the family to which the ant thrushes and ant pittas belong (Formicariidae), are almost completely nonmigratory. Others, such as the New World and Old World warbler families (Parulidae and Sylviidae, respectively), contain some species that are permanent residents, others partial migrants, and still others long-distance temperate–tropical migrants. In most families of temperate zone songbirds, such as the titmice and chickadees (Paridae), some species are at least partial migrants, that is, with some populations that consist partly or largely of migratory individuals. Almost all seabirds are migratory, in the sense of spending nonbreeding periods at sea at feeding areas substantially distant from their nesting areas. Waterfowl that breed in the temperate zones or the Arctic are almost all migratory to varying degrees. Shorebirds breeding at high latitudes are nearly all long-distance migrants. Falcons and hawks that breed in the temperate zones or the Arctic are also mostly migratory—some only short-distance migrants but others showing

TABLE 1.1. Migratory bird species summarized by order.

| Orders | Families | Total Species | Migratory Species |
|---|---|---|---|
| Tinamous (Tinamiformes) | 1 | 47 | 0 |
| Rheas (Rheiformes) | 1 | 2 | 0 |
| Ostrich (Struthioniformes) | 1 | 1 | 1 |
| Emu and Cassowaries (Casuariiformes) | 2 | 4 | 2 |
| Kiwis (Dinornithiformes) | 1 | 5 | 0 |
| Loons (Gaviiformes) | 1 | 5 | 5 |
| Grebes (Podicipediformes) | 1 | 19 | 15 |
| Penguins (Sphenisciformes) | 1 | 17 | 15 |
| Petrels and Albatrosses (Procellariiformes) | 4 | 114 | 114 |
| Pelicans and Relatives (Pelecaniformes) | 4 | 67 | 48 |
| Ducks and Geese (Anseriformes) | 2 | 162 | 124 |
| Flamingos (Phoenicopteriformes) | 1 | 6 | 6 |
| Wading Birds (Ciconiiformes) | 5 | 117 | 54 |
| Raptors (Falconiformes) | 5 | 313 | 108 |
| Quail and Relatives (Galliformes) | 6 | 284 | 13 |
| Cranes and Relatives (Gruiformes) | 12 | 207 | 74 |
| Shorebirds (Charadriiformes) | 19 | 356 | 274 |
| Sandgrouse (Pterocliformes) | 1 | 16 | 11 |
| Pigeons and Relatives (Columbiformes) | 1 | 308 | 27 |
| Parrots and Relatives (Psittaciformes) | 3 | 368 | 37 |
| Mousebirds (Coliiformes) | 1 | 6 | 0 |
| Cuckoos and Turacos (Cuculiformes) | 3 | 164 | 50 |
| Owls (Strigiformes) | 2 | 215 | 18 |
| Nightjars and Relatives (Caprimulgiformes) | 5 | 120 | 35 |
| Swifts and Hummingbirds (Apodiformes) | 3 | 443 | 89 |
| Trogons (Trogoniformes) | 1 | 40 | 5 |
| Motmots and Relatives (Coraciiformes) | 10 | 219 | 48 |
| Woodpeckers (Piciformes) | 6 | 412 | 27 |
| Perching Birds (Passeriformes) | 98 | 5893 | 1400 |
| Total | 204 | 9930 | 2600 |

intercontinental movements. Owls of these same latitudes, on the other hand, are in some cases migrants, in other cases permanent residents.

## Migratory Birds: Geographical Patterns

The relative abundance of migratory species also differs markedly among major geographical areas (Table 1.2). The large landmasses of the North

TABLE 1.2. Living resident and migratory nonoceanic bird species in various world regions.

| Region | Total Species | Introduced Species | Permanent Residents | Intraregional Migrants (Altitudinal) | Interregional Migrants: Breeding | Interregional Migrants: Nonbreeding |
|---|---|---|---|---|---|---|
| North American | 1825 | 40 | 967 | 585 (156+) | 126 | 147 |
| Nearctic | | | | 410 (57+) | 13 | |
| Nearctic-Neotropical | | | | 76 | 108 | |
| Neotropical | | | | 99 (99+) | 5 | |
| South American | 3260 | 15 | 2881 | 266 (16+) | | 113 |
| European | 759 | 17 | 223 | 116 | 185 | 235 |
| Intra-European | | | | 116 (11+) | | |
| European–African | | | | | 185 | |
| Asian | 2588 | 5 | 1686 | 827 (182+) | 46 | 29 |
| African | 2354 | 16 | 1724 | 411 (51+) | | 219 |
| Intratropical | | | | 277 (44+) | | |
| Southern African | | | | 134 (7) | | |
| Australasian | 1438 | 26 | 931 | 272 (51) | | 235 |
| Pacific Islands | 494 | 59 | 305 | 2 (2) | | 187 |
| Antarctica | 70 | 2 | 15 | | | 55 |

Principal Sources:

Clements, J. F. 2007. *The Clements Checklist of Birds of the World*, 6th ed. Comstock Publishing Associates (Cornell University Press), Ithaca, NY.
AOU checklist of North American birds. http://www.aou.org/checklist/north/index.php.
Avibase. Bird checklists of the world. http://avibase.bsc-eoc.org/checklist.jsp?lang = EN.
Oriental Bird Club. http://www.orientalbirdclub.org/publications/checklist/obcchecklist.txt.
*The Birds of Africa*. Vol. 1–7. Academic Press, New York.

Temperate and Arctic zones contain many migrant species, most of which breed in the region. In tropical and subtropical regions of the continents, wintering migrants are numerous. Tropical island archipelagos such as the East and West Indies are also wintering areas for many migrants, as are smaller island areas throughout the Atlantic, Pacific, and Indian oceans. Many of the migrants to distant oceanic areas are nonbreeding shorebirds and seabirds. South Temperate Zone regions are also home to many breeding migrant species, and the birds that breed in the Antarctic are almost all migratory. Intratropical migration patterns are well developed in Africa, and to a lesser extent in Australia and the New World. Many birds of mountainous regions, regardless of latitude, show altitudinal movements.

Several major migration systems can be recognized in different world regions. In the New World, these include the Nearctic, Nearctic–Neotropical, Neotropical, and South American migration systems. The Nearctic migration system comprises land and freshwater birds that breed in Canada and the United States and winter primarily north of central Mexico. This migration complex involves about 423 species, including many species of waterfowl and other freshwater birds, hawks and owls, and small land birds. About 13 of the Nearctic species that breed at high latitudes migrate to Asia rather than to more-southern parts of North America.

The Nearctic–Neotropical migration system includes bird species that breed in North America and winter in southern Mexico, Central America, the West Indies, and South America. About 184 species of land, freshwater, and coastal marine birds are involved. About 76 of these species spend the nonbreeding season in Mexico, the West Indies, or Central America, but 108 species extend their winter ranges into South America. The Neotropical migration system comprises about 104 species, most of which are altitudinal migrants. About 5 species breed in the North American tropics and winter in South America. We know more about patterns of altitudinal migration in Mexico and Central America than anywhere else, and what we see here suggests that we have much to learn about altitudinal migration in mountainous regions elsewhere in the world.

The South American migration system comprises about 266 land, freshwater, and coastal marine birds. Most of these species breed in the temperate region of southern South America and migrate north in the austral winter, but at least 31 species show short-distance or altitudinal movements in the tropics. About 77 of the species breeding in temperate South America are fully migratory. Some 44 species, mostly tyrannid flycatchers and swallows, winter in the humid tropical region centered on the Amazon basin. For more than two-thirds of these species, ranges of nonbreeding migrants overlap ranges of residents of the same species.

The Old World exhibits European, European–African, and Asian migration systems. In Europe, about 116 of 524 breeding species of land, freshwater, and coastal marine birds are short-distance migrants that winter within the region. The European–African migration system includes 185 species of land, freshwater, and shorebirds of thirty-two families that winter in Africa. These species are primarily hawks and falcons, waterfowl, shorebirds, Old World warblers, thrushes, swallows, pipits, wagtails, and shrikes. Of these, 62 species, particularly many hawks and falcons, shorebirds, Old World warblers, thrushes, and shrikes, winter exclusively in sub-Saharan Africa. Of the remaining species, some winter in Africa and some in areas of the Middle East.

The rich Asian migration system involves birds that move from breeding areas in eastern Eurasia to wintering areas in southern and southeastern Asia, the Philippines, and the East Indies. About 827 species of forty-four families show migratory movements within this region. About 35 species of sandpipers and plovers, 1 tern, and 10 species of land birds continue farther south to winter in the Australo–Papuan region. In Russia, China, and other parts of eastern Asia, at least 66 species of thrushes, flycatchers, Old World warblers, finches, and other passerines are short-distance or altitudinal migrants that do not reach the tropics.

Africa exhibits two well-defined migration systems. The southern African migration system involves 134 fully or partially migratory land and freshwater birds that breed in southern Africa and winter farther north. Waterfowl, rails, herons, kingfishers, cuckoos, swifts, swallows, and a variety of passerine birds are prominent in this system. Many birds breeding in Europe and Asia appear in southern Africa during the Northern Hemisphere winter. These include many species of sandpipers and plovers, gulls and terns, hawks, and Old World warblers.

The intratropical African migration system is centered on the equatorial region and the semiarid belts to the north and south. About 277 species, ranging from herons and plovers to larks and finches, move primarily north and south in this region, tracking favorable conditions related to wet and dry season weather patterns. Some cuckoos breed in the semiarid zones both north and south of the equator, switching places through migration when not breeding. In more-arid regions, movements of some species are best described as nomadic. Altitudinal migrants are frequent in the mountains of eastern Africa.

The Australo–Papuan region, comprising Australia and New Guinea, holds a largely self-contained migration system. This migration system consists of about 272 land and freshwater birds. Some are migrants between Australia and New Guinea, others move between Tasmania and mainland

Australia, and still others show latitudinal and altitudinal movements within mainland Australia. In addition, many land birds and waterbirds perform irregular movements within interior mainland Australia. Some 169 species show partial migration patterns in Australia and Tasmania. Among Tasmanian birds, only 20 species are migratory, and only 4 are fully migratory, completely leaving the island for the Australian mainland.

Northern Hemisphere migrants also reach many Pacific islands, including New Zealand, although few species are involved. Long-distance migrants such as plovers and sandpipers most frequently visit these isolated regions.

Seabird migration systems relate less to latitudinal temperature patterns and more to the locations of suitable nesting areas and productive ocean waters. About 234 species of birds of eighteen families are largely or entirely pelagic in their distributions in the nonbreeding season (Table 1.3). Only very general migration patterns are apparent. In the eastern North Atlantic, about 31 species of loons, petrels, shearwaters, gannets, cormorants, sea ducks, skuas, gulls, and alcids breed in Arctic and subarctic areas and winter in pelagic or offshore coastal areas. In the western Atlantic, about 28 of the species of these groups breed in northern areas and winter in waters off the coasts of Canada and the United States. The North Pacific has a richer fauna. About 53 species of these seabirds winter southward through the Aleutian Islands and along the eastern and western coasts of the Pacific. Tropical and subtropical oceans host the richest and most diverse fauna of pelagic birds, including numerous petrels, shearwaters, terns, boobies, tropicbirds, cormorants, and frigatebirds.

In the southern oceans surrounding Antarctica, at least 94 species of penguins, albatrosses, petrels, cormorants, skuas, and related birds breed and forage to considerable distances from breeding localities. In tropical and warm temperate ocean areas, at least 97 species of seabirds of eleven families breed and wander widely over ocean areas distant from their nesting islands or coastal rookeries.

## The Challenge of Climatic Change

Migratory birds face major challenges of survival in the face of rapid, human-induced global change. Over the past century, the earth's climate has warmed by 0.8°C, and by AD 2100, warming will likely be between 2.0°C and 4.5°C. Because they depend on habitats and resources in different areas at different stages of the annual cycle, populations of seasonal migrants in

TABLE 1.3. World distribution of migratory oceanic birds.[1]

| | | Intraregional Migrants | | | | | Interregional Migrants | | | | |
| | | | | | | | Breed: Antarctic/Southern | | Breed: Tropics | Breed: Arctic/Subarctic | |
| Taxonomic Group | Total Species | Antarctic/ Southern | Southern/ Subtropical | Subtropical/ Tropical | Subtropical/ Northern | Northern/ Arctic | Winter to Temperate Zone or Tropics | Winter to Northern Hemisphere | Winter to Northern Hemisphere | Winter to Temperate Zone or Tropics | Winter to Southern Hemisphere |
|---|---|---|---|---|---|---|---|---|---|---|---|
| Penguins | 15 | 13 | 1 | 1 | | | | | | | |
| Loons | 1 | | | | | | | | | | |
| Albatrosses | 13 | 6 | 9 | 1 | | | | | 3 | 1 | 3 |
| Petrels and shearwaters | 77 | 6 | 14.5 | 22 | 5.5 | 1 | 10 | 14 | 1 | | |
| Storm-Petrels | 20 | 1 | 1.5 | 9.5 | 3 | | 2 | 1 | | 2 | 1 |
| Diving-Petrels | 4 | 1 | 2 | 1 | | | | | | | |
| Tropicbirds | 3 | | | 3 | | | | | | | |
| Pelicans | 2 | | | 2 | | | | | | | |
| Boobies and gannets | 10 | | 1 | 7 | | | 1 | | | 1 | |
| Cormorants | 21 | 7 | 7 | 2 | 3 | 2 | | | | | |
| Frigatebirds | 5 | | | 5 | | | | | | | |
| Sea ducks | 9 | | | | | 4 | | | | 5 | |
| Phalaropes | 2 | | | | | | | | | | |
| Sheathbills | 2 | 2 | | | | | | | | | 2 |

TABLE 1.3. Continued

| | | Intraregional Migrants | | | | | Interregional Migrants | | | | |
| | | | | | | | Breed: Antarctic/Southern | | Breed: Tropics | Breed: Arctic/Subarctic | |
| Taxonomic Group | Total Species | Antarctic/ Southern | Southern/ Subtropical | Subtropical/ Tropical | Subtropical/ Northern | Northern/ Arctic | Winter to Temperate Zone or Tropics | Winter to Northern Hemisphere | Winter to Northern Hemisphere | Winter to Temperate Zone or Tropics | Winter to Southern Hemisphere |
|---|---|---|---|---|---|---|---|---|---|---|---|
| Skuas and jaegers | 7 | | | | | | 2 | 1 | | | 4 |
| Gulls | 7 | | | 1 | 1 | 2 | | | | 2 | 1 |
| Terns | 14 | | 2 | 11 | | | | | | | 1 |
| Auks, murres, puffins | 22 | | | | 3 | 13 | | | | 6 | |
| Total | 234 | 29 | 38 | 65.5 | 15.5 | 22 | 15 | 16 | 4 | 17 | 12 |

¹Zonal separations roughly correspond to Arctic and Antarctic circles and Tropics of Cancer and Capricorn. Some species are partitioned between two zones, giving rise to the 0.5 values.

every world region will be affected by climatic change. The areas that migrants use seasonally include their breeding ranges, staging and stopover locations during migration, and areas occupied during the nonbreeding period. In one sense, because of their specialization for use of different habitats or geographical areas at different times, migratory species might seem to be at greater risk of extinction than permanent residents. A change in any one of the areas used during the annual cycle might cause their evolutionary strategy to fail. Some ornithologists conclude that such dependence on multiple geographical areas places migratory species at greater risk than resident species in the face of global climate change.

On the other hand, migratory capability is an extension of basic physiological and behavioral adaptations for local movements, homing, and the annual reproductive cycle, and it must possess a degree of flexibility. Climatic changes over long geological time have tested the ability of migrants to adjust breeding and nonbreeding ranges and alter migration routes. Perhaps this adaptability is adequate to respond to the rapid environmental changes now occurring. Or, of course, some migratory birds may be able to adjust quickly, while others may not.

Several factors could constrain the ability of migratory birds to respond to changing climate. Lack of genetic variability or phenotypic plasticity might limit the capacity of species to respond to changing conditions. Limited dispersal ability might also slow the ability of the species to shift their geographical ranges in response to changing conditions. An increase in frequency of extreme weather events, overall loss or degradation of one or more of the habitats on which they depend throughout the annual cycle, and changed influences of competitor species, predators, or diseases may also impact migratory species negatively.

I suspect that population limitation of migratory species is a dynamic relationship involving both breeding and nonbreeding areas. If conditions in the wintering range favor increased survival, expansion of the breeding range into regions with lower reproductive success will tend to occur. If conditions in the breeding range favor increased reproductive success, expansion of the nonbreeding range into regions with lower survival will likely result. Inasmuch as breeding success and nonbreeding survival vary from year to year, the outcome of this relationship changes through time. Both human modification of the landscape and global climatic change obviously influence this process, as well.

Issues relating to migratory birds attract enormous scientific concern. The Smithsonian Institution sponsored symposia on migratory birds in the New World in 1977 and 1989. In 2002, another Smithsonian symposium, entitled Birds of Two Worlds: Advances in the Ecology and Evolution of Temperate–Tropical Migration Systems, considered migration patterns worldwide. The first organized effort to focus directly on climate issues was a symposium, Bird Migration in Relation to Global Change, held at the University of Constance, Germany, in March 2003. Most recently, a congress on Bird Migration and Global Change took place at Algeciras, Spain, in March 2007.

Since 2006, the United Nations Environment Programme has supported World Migratory Bird Day (WMBD), a global initiative focused on migratory birds and their conservation. The secretariat of the African–Eurasian Migratory Waterbird Agreement initiated WMBD, which now involves several other cooperating organizations. Actually a 2-day event on the second weekend in May, each annual program emphasizes a particular issue relating to migratory birds. In 2009 WMBD focused on man-made structures that pose dangers to migrating birds.

## Patterns of Response to Climate Change

Changing climates are altering migratory behavior in several ways. These include (1) change in the relative numbers of migrant and resident individuals in local populations, (2) change in the distance between areas occupied at different stages in the annual cycle, (3) change in the direction of migratory movements, and (4) change in timing and speed of migratory movements. Some of these responses appear to be simple phenotypic responses of individuals to altered environmental conditions, whereas others are genotypic or microevolutionary responses.

Changes in the proportion of migrants and residents can occur if ameliorated climatic conditions during the nonbreeding season in a particular location favor more individuals' becoming permanent residents. Or, if weather conditions and resource availability patterns become more strongly seasonal, an increase in the migratory fraction of the population can occur. Since patterns of global climatic change vary from region to region, both of these changes are expected.

With climatic warming, favorable habitats for breeding by many species may appear at higher latitudes. This may lead to longer migrations by individuals colonizing such areas, if the areas suitable during the nonbreeding season remain the same. Or, as in the case of leapfrog migration patterns, individuals breeding farther poleward may tend to move to nonbreeding areas even farther from their breeding areas. Lengthened migrations could be a severe challenge if stopover sites also became less frequent or less favorable. On the other hand, amelioration of climate in nonbreeding areas might allow nonbreeding birds to remain closer to their breeding ranges.

Since climatic changes are uneven across large continental masses, changes in the direction of migratory movements are likely to occur. Regions suitable for occupation during breeding or nonbreeding periods might appear at higher latitudes but at longitudes different from previously suitable regions. Likewise, previously favorable breeding or nonbreeding areas might shift in longitude.

With climatic change, the seasons favorable for occupation of the breeding range or occupation of the nonbreeding areas may lengthen or shorten, leading to a need for changes in the migration schedule. Opposing selection pressures at different seasons may slow or prevent adjustments in timing of migration, and life history features may fall out of synchrony with seasonal progression. Especially for long-distance migrants, such as Eurasian birds wintering in sub-Saharan Africa or Nearctic migrants wintering in the tropics, correlations between habitat conditions on breeding and nonbreeding ranges may be very weak or even lacking. Endogenous factors and photoperiodic responses that are not directly tied to temperature are what control the annual cycles of most of these species. As a result, many birds may fail to adjust migration schedules to take advantage of optimal conditions for breeding.

A central issue of climatic change for all migratory species is their degree of vulnerability to extinction. Migratory birds have dealt successfully with major climatic shifts during periods of Pleistocene glaciation, although some extinctions almost certainly did occur. Now, however, climate is changing more rapidly than during the Pleistocene. The extensive transformation and fragmentation of habitats by human activities may constrain the ability of many species to respond to geographical shifts in their optimal habitat. For birds and other land animals that occupy habitats defined by plant community structure, the ability of plant species to move in response to changing climate may be critical. In addition, even though birds are highly mobile, strong site fidelity and the mechanism by which site fidelity is established in juveniles might also limit the capacity of migratory birds to occupy new regions of suitable habitat.

Short- and long-distance migrants might also be differentially vulnerable to extinction in the face of changing climates. Short-distance migrants tend to respond directly to immediate weather conditions within the local environment. Inasmuch as the weather conditions on their nonbreeding ranges are likely to be correlated with those on their nearby breeding ranges, appropriate changes in migratory schedules are quite likely. Long-distance migrants must respond to environmental cues that are correlated with changes on their distant breeding or nonbreeding areas and along their migration routes. How quickly they can respond is uncertain.

In the following chapters we shall first explore the nature of the climatic and environmental changes that are occurring. We shall then examine responses by birds of various world regions and ecological groups. Finally, we shall evaluate the capacity of migratory birds to adapt to the changes in climate that are foreseen.

# Summary

Migratory birds throughout the world are facing climatic change. Over 26 percent of all bird species show some pattern of migration. Major systems of migration by land and freshwater birds exist in North America, Eurasia, and the continents of the Southern Hemisphere. In the Northern Hemisphere these include patterns of short-distance, intracontinental movement and long-distance temperate–tropical or intercontinental movements. Seabirds at all latitudes show latitudinal movements between breeding and nonbreeding areas or nonbreeding dispersal to distant ocean areas. The capacity of these diverse migrants to adjust to changing climates and food supplies will determine whether or not migration will remain a major phenomenon of avian ecology.

## KEY REFERENCES

Alerstam, T., A. Hedenstron, and Susanne Äkesson. 2003. "Long-distance migration: Evolution and determinants." *Oikos* 103:247–260.

Chesser, R. T. 1994. "Migration in South America: An overview of the austral system." *Bird Conservation International* 4:91–107.

Cox, G. W. 1985. "The evolution of avian migration systems between temperate and tropical regions of the New World." *American Naturalist* 126:451–474.

Dingle, H. 2004. "The Australo–Papuan bird migration system: Another consequence of Wallace's line." *Emu* 104:95–108.

Greenberg, R. and P. Marra (Eds.). 2005. *Birds of Two Worlds: The Ecology and Evolution of Migration.* Johns Hopkins University Press, Baltimore.

Hockey, P. A. R. 2000. "Patterns and correlates of bird migration in sub-Saharan Africa." *Emu* 100:401–417.

Huntley, B., Y. C. Collingham, R. E. Green, G. M. Hilton, C. Rahbek, and S. G. Willis. 2006. "Potential impacts of climatic change upon geographical distributions of birds." *Ibis* 148:8–28.

Joseph, L. 1996. "Preliminary climatic overview of migration patterns in South American austral migrant passerines." *Ecotropica* 2:185–193.

Moreau, R. E. 1972. *The Palearctic–African Bird Migration System*. Academic Press, New York.

Rappole, J. H. 1995. *The Ecology of Migrant Birds: A Neotropical Perspective*. Smithsonian Institution Press, Washington, DC.

# PART II

## The Changing Environment

Human activities are modifying the chemistry and physics of the earth's atmosphere, as well as characteristics of the land surface, freshwater lakes and streams, and the oceans. These changes are altering the earth's energy balance, its basic patterns of atmospheric and ocean circulation, and climatic conditions of all of the earth's ecosystems. In turn, these effects are altering the habitats and resources that are critical for migratory birds. Direct human impacts on the earth's ecosystems are interacting with these relationships, creating a complex pattern of global environmental change.

# Chapter 2

# *Global Climate Change*

Climates are changing worldwide, but nowhere more than at high latitudes. In mid July 1977, my wife and I spent a week at Kivalina, Alaska, an Inupiaq Native village located on a barrier island on the Arctic Ocean above the Arctic Circle. At that midsummer time we could see the sea ice a mile or so off-shore, and hunters were still taking seals on the ice. Since then, retreat of the ice pack has made summer hunting on the ice impossible and has left the Kivalina shore exposed to stronger wave action and erosion. A seawall of sand-bags was constructed in 2006 at a cost of $3 million, but maintaining this defense is difficult. Moving the village to a safe site would likely cost $155 million. This is not an unusual story for regions at high northern latitudes. Changing climate is affecting the lives of people and the birds, mammals, fish, and plants they traditionally harvest all around the Arctic Ocean.

In this chapter we examine the basic patterns of climate change in terrestrial, freshwater, and marine environments. These changes drive the worldwide patterns of alteration of migratory bird habitats that we shall consider in detail in the next chapter.

## Altered Global Radiation Balance

The addition of greenhouse gases to the atmosphere is the strongest influence of human activities on climate. The principal greenhouse gases include

21

water vapor ($H_2O$), carbon dioxide ($CO_2$), methane ($CH_4$), nitrous oxide ($N_2O$), ozone ($O_3$), chlorofluorocarbons (CFCs), and other halogen compounds. All of these have increased in concentration in the atmosphere directly or indirectly because of human activities. All act to trap heat radiation in the lower atmosphere and increase the temperature of the earth's surface.

The two most influential greenhouse gases are carbon dioxide ($CO_2$) and methane. Atmospheric $CO_2$ has risen from about 280 parts per million (ppm) in about AD 1750 to 387 ppm in early 2009. The annual rate of increase has accelerated, rising from 1.49 ppm during the 1990s to 1.93 ppm in 2000–2006. The rise in $CO_2$ is due primarily to the burning of fossil fuels, combined with release of $CO_2$ in cement production and through deforestation and other reductions of plant biomass. In addition, since 2000, the fraction of released $CO_2$ that is absorbed in land and water systems has decreased. Warming of the Arctic may lead to even larger releases of $CO_2$ due to decomposition of organic matter in thawing permafrost zones. Carbon dioxide is the most important of the gases for future greenhouse warming because of the difficulty of halting its addition to the atmosphere.

Atmospheric methane has risen from about 700 parts per billion to over 1774 parts per billion since about AD 1750. Methane is released in petroleum production and biomass burning, as well as from ruminant livestock, rice paddies and other shallow waters, warming marine sediments, and thawing permafrost in Arctic regions. Concentrations of $CO_2$ and methane now far exceed their natural levels over the past 650,000 years.

Other greenhouse gases have also increased. Nitrous oxide has risen by about 15 percent, mainly because of the action of denitrifying bacteria on nitrogen fertilizers and natural nitrogen compounds in anaerobic agricultural soils. Since the 1950s, atmospheric concentrations of CFCs and other halogens, from release of industrial compounds and burning of biomass, have also increased. Ozone has increased in the lower atmosphere, largely as a photochemical product of pollutants released by internal combustion engines.

For several greenhouse gases, additions will persist in the atmosphere for a very long time, so their influence is not easily counteracted. For $CO_2$, even complete cessation of input to the atmosphere will lead to retention of about 40 percent of the augmented input after 1000 years. Others, particularly nitrous oxide and CFCs, have persistence times of over a century.

Human activity has altered atmospheric chemistry in still other ways, including weakening the stratospheric ozone layer. At an elevation of 15–35 kilometers, ozone, formed by the action of ultraviolet radiation on oxygen, creates the "ozone shield" that absorbs most ultraviolet radiation and protects terrestrial life. Ultraviolet radiation in certain wavelengths does reach the earth's surface. In the stratosphere, halogen compounds contribute to the breakdown of ozone, especially in polar regions. This breakdown allows increased amounts of biologically active ultraviolet radiation to reach the earth's surface. In 2006 the ozone depletion "hole" centered over Antarctica was the largest ever recorded, peaking at an area of 29.5 million square kilometers in early September. In 2008, the ozone hole was only slightly smaller. Although international efforts are reducing halogen releases, the quantities now present will affect ozone levels for many decades. In addition, stratospheric conditions over the northern polar region have shown a tendency to become similar to those of the Antarctic, with an increase in frequency of stratospheric low ozone events in November and December.

Some of these changes in atmospheric chemistry induce a positive feedback. The greenhouse warming that has already occurred in the Arctic is an example. With the accelerated melting of land and sea ice, which reflects much incoming solar radiation, more solar radiation is absorbed, intensifying the warming effect. With warming of the tundra environment, the spread of trees and shrubs also increases the radiation absorption of the land surface. To make things even worse, Arctic soils contain immense amounts of carbon in the form of peat, so Arctic warming is likely to cause large additional releases of $CO_2$ by increasing decomposition. Climatic warming also means that northern wetlands will likewise become major sources of additional $CO_2$ and methane. In fact, the North American Arctic tundra appears to have changed from a net storage system for carbon to a net source.

The influence of many of these changes is greatest at high latitudes. Arctic and Antarctic regions, home to hundreds of species of migratory marine, freshwater, and land birds, are now feeling some of the strongest effects of human alteration of the atmosphere.

## Extent of Global Climatic Change

According to the latest report of the Intergovernmental Panel on Climate Change, the evidence for global climatic warming is unequivocal. Over the century ending in 2005, mean global temperature increased about 0.74°C, a value significantly greater than the former estimate of 0.6°C for the period 1901–2000. Mean global temperatures for 11 of the 12 years from 1995 to 2006 are among the warmest years noted since about 1850.

The timing of seasons and prevailing seasonal temperatures are also changing over the breeding and nonbreeding ranges of hundreds of migratory birds. Over the past 54 years, the seasonal cycle of temperature over land areas has become about 1.7 days earlier. Similar shifts have occurred over some ocean areas, such as the eastern North Pacific and southern North Atlantic. In other ocean areas, including the interior North Pacific and the northern North Atlantic, the seasonal cycle has become somewhat later. In addition, some continental areas, including interior Asia and North America, now experience warmer winters, leading to smaller differences between summer and winter temperatures. Many other areas are seeing warmer summers.

The increase in global temperature is leading to increased storage of heat in the oceans. Since the 1950s, the increase in heat content of the oceans has been almost three times that for the atmosphere. Increased heating of the oceans has decreased areas of sea ice and thinned the perennial Arctic ice pack. Sea ice conditions, in both the Arctic and Antarctic, are critical for foraging of many sea birds.

Global warming has also affected the hydrologic cycle. Satellite data suggest that both the quantity of water in the atmosphere and total precipitation increase about 7 percent for each 1-kelvin increase in global temperature. Warming has also led to a decrease in the masses of continental ice caps and mountain glaciers.

## Altered Regional Climate Dynamics

Humans have initiated changes in the global climate system that will continue for centuries, even if no further increase in greenhouse gases occurs.

### *Future Climate Change Commitment*

We are committed to greater changes in global climate than we have so far seen. Ocean waters will continue to warm, but the thermal inertia of the oceans is so great that an equilibrium between the heat content of oceans and atmosphere will not develop for decades or centuries. Climate simulations suggest that even if atmospheric composition remains at it is today, mean global temperature will rise by more than 1°C by AD 2400. Furthermore, global temperature is not predicted to decline significantly over the next millennium.

If atmospheric composition continues to change at the current rate, much greater changes in global temperature and sea level are certain (Figure 2.1). Global temperature increase will likely equal 2–6°C by AD 2400, depending on the level at which increase in $CO_2$ is stabilized. In addition, sea level rise, due mainly to thermal expansion of warmer sea water, will be about 10 centimeters per century through AD 2400 if atmospheric composition remains at present levels, and about 25 centimeters per century if present emission trends continue. Ultimately, thermal expansion alone could increase sea level 0.5–2.0 meters. Disappearance of glaciers and small ice caps would further raise sea level 0.2–0.8 meter. Even greater sea level rise would result from melting of Antarctic and Greenland ice caps. Collapse of the West Antarctic Ice Sheet alone would lead to catastrophic sea level rise in midlatitude and northern oceans. Some projections of sea level rise in this event equal 6.3 meters.

The biota of the world's oceans will be profoundly affected by future climate change. Changes in temperature, current patterns, and sea level will greatly alter the feeding and nesting areas of seabirds, almost all of which are migratory to some degree.

### Oceanic and Atmospheric Interactions

Warming of the atmosphere and oceans is altering large-scale climatic systems. Oceanic regions are centers of cycles of atmospheric and oceanic circulation that vary in length from years to decades. These cycles result from the modification of basic atmospheric processes by the irregular boundaries of continents and oceans and by the influence of mountain systems. Their influence, centered on ocean areas, carries over to the adjacent continents. They also complicate our efforts to evaluate how regional climates are being affected by global change.

Several major oceanic–atmospheric cycles are now recognized. Principal among these are the North Atlantic Oscillation (NAO), the El Niño Southern Oscillation (ENSO), the Pacific Decadal Oscillation (PDO), and the Antarctic Oscillation (AAO).

The NAO dominates winter weather patterns from central North America to Europe and much of northern Asia and has a strong influence on wintering patterns of many short-distance migrants. The NAO is a shift in the relationship between a subtropical Atlantic area of high atmospheric pressure, centered near the Azores, and an Arctic low, centered near Iceland. This relationship tends to remain in one phase for several years,

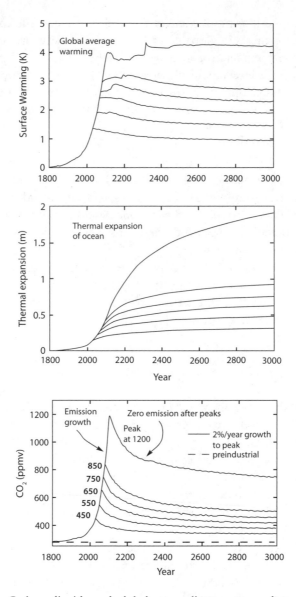

FIGURE 2.1. Carbon dioxide and global mean climate system changes projected through AD 3000. Top: Decline in atmospheric carbon dioxide concentration following a complete cessation of input at various concentrations. Middle: Surface warming (in kelvins) corresponding to peak carbon dioxide levels in top panel. Bottom: Sea level rise due to thermal expansion for warming scenarios in middle panel. (From Solomon, S., G.-K. Plattner, R. Knutti, and P. Friedlingstein. 2009. "Irreversible climate change due to carbon dioxide emissions." *Proceedings of the National Academy of Sciences USA* 106:1705.)

although phase changes tend to be irregular. The NAO influences the track of weather systems between about 40° and 60° N. In its positive phase, the subtropical high-pressure center tends to be stronger than average and the Icelandic low deeper than average. During this phase, strong westerly winds drive more strong winter storms across the Atlantic on a more northerly track. The result is warm and wet winters in Europe and cold and dry winters across northern Canada and Greenland. The eastern United States experiences mild, wet winter conditions. During the negative phase, the subtropical high is weak and the Icelandic low is also weak. This reduces the pressure gradient, leading to weaker westerly winds that drive fewer weak winter storms across the Atlantic on a more southern pathway. This results in a flow of moist air over the Mediterranean region and cold, dry air over northern Europe. On the east coast of North America, the result is colder, snowier winter weather. Since the mid-1990s, the NAO has tended to remain in the positive phase with greater frequency.

The ENSO is a combined pattern of atmospheric behavior, the Southern Oscillation (SO), and of the ocean's response, the cycle of El Niño (EN) and La Niña conditions in the tropical Pacific. The Southern Oscillation is an alternating pattern of change in atmospheric pressures between the eastern and western tropical Pacific Ocean. El Niño, referring to the Christ child, was first applied to warm ocean conditions that appeared at intervals of several years along the coast of Peru, usually around Christmas. La Niña is the name more recently given to the opposite set of ocean conditions.

In detail, ENSO shows the following pattern: In the tropical Pacific, during La Niña years, strong trade winds drive equatorial currents west from the coast of South America, leading to upwelling of cold, nutrient-rich water that supports the high productivity of the coastal waters. At the peak of this process, warm surface water thus becomes concentrated in the Australian–Indonesian region, accompanied by a strong tropical low-pressure system with heavy convectional rainfall. Eventually, this tropical low spreads eastward, reducing the strength of the trade winds and allowing ocean currents to carry warm water eastward. When these currents reach South America, they seal off the upwelling along the coast, creating an El Niño. During El Niños, coastal areas of South America receive heavy rains, often resulting in heavy flooding. Suppression of upwelling catastrophically reduces the productivity of fish in coastal waters, and populations of boobies, cormorants, pelicans, penguins, and other seabirds are decimated. At the same time, drought conditions tend to occur in the western Pacific region.

ENSO strongly influences weather almost worldwide, particularly by contributing to droughts in Asia, Africa, Australia, and Central America. In North America, El Niños tend to push winter storms across the southern United States but create less storminess and milder conditions across the North. El Niños tend to promote warm, dry winters in the Pacific Northwest and cool, wet winters in the American Southwest. La Niñas promote colder and stormier than average conditions in the northern United States and warmer and less stormy conditions in the South. These patterns also influence the wintering distributions of many migratory land and freshwater birds.

The ENSO oscillation has tended to vary in period from 2 to 7 years, but the interval between El Niños has decreased in recent years and La Niñas have become weaker. Since 1970, El Niños have occurred every 2.2 years. Some scientists strongly suspect that increased solar heating in the tropics is contributing to this pattern.

The Pacific Decadal Oscillation (PDO), a pattern of climate variability over the North Pacific, tends to show a periodicity of 20–30 years. In the warm phase of the cycle, waters in the western North Pacific are cooler, and those in the equatorial Pacific are warmer than average. During the cool phase, these patterns switch. Two full PDO cycles have occurred over the past century. A cool PDO regime existed from 1890 to 1924 and again from 1947 to 1976, whereas a warm PDO regime prevailed from 1925 to 1946 and from 1977 until 2007. A shift to a cool phase may now be occurring. During the warm phase, marine productivity is high in Alaskan waters and low along the west coast of the United States. During the cold phase, the opposite pattern of marine productivity prevails. The exact cause of PDO cycles and whether or not this pattern is affected by global warming are uncertain, but the changes in marine productivity are a significant influence on populations of many North Pacific seabirds, ranging from non-breeding shearwaters and petrels to many breeding alcids, petrels, gulls, and cormorants.

The PDO is also linked to climatic patterns, particularly rainfall, in North America. The warm phase tends to promote warm, dry conditions in the Pacific Northwest and cool, wet conditions in the Southwest. The PDO interacts with ENSO in complex fashion because of their different periods.

In the Southern Hemisphere, the Antarctic Oscillation (AAO) is an alternation in the gradient of atmospheric pressure between a polar zone of high pressure and subpolar low-pressure zones. This difference influences the strength of westerly circumpolar flow of air masses around Antarctica, which has tended to increase in recent decades. This change appears to be related to stratospheric ozone depletion and to contribute to the warming of the Antarctic Peninsula and southernmost South America.

## Climate Change in Terrestrial Environments

Climatic warming is greatest over the continents and particularly in the Northern Hemisphere. Over the last half century, the Northern Hemisphere was warmer than at any time during the past 1300 years. North of latitude 45°, the growing season appears to have lengthened by 12 to 19 days. The potential breeding season for many migratory birds thus is substantially longer.

At the highest northern latitudes, climatic warming is much greater than the global average. In Alaska, western Canada, and eastern Russia, mean temperature has increased as much as 3–4°C in the last half century, more than twice the global average. The rate of summer warming has increased in both Alaska and western Canada and is now 0.3–0.4°C per decade, more than twice the rate during 1961–1990.

High-latitude warming has altered ecosystem dynamics. Spring thawing of the Arctic tundra areas of North America and Eurasia has advanced 2.0–3.3 days per decade since the 1960s. On the northern Alaskan coast, the snow-free season has lengthened by about 2.5 days per decade since the 1960s, with greater lengthening inland. Across boreal Eurasia, satellite data show an advance in spring greening of 7.1 days over the period 1982–1999. Conditions suitable for breeding by many long-distance migrant land and freshwater birds thus appear substantially earlier in summer. Recent studies also show an increase in precipitation at high latitudes, contributing to increased river flow into the Arctic Ocean.

Climatic warming is also substantial farther south in Eurasia and North America. Europe in general has warmed about 1°C, slightly more than the world at large. Warming is greatest in the Mediterranean region, northeastern Europe, and mountain regions. In Spain, the growing season lengthened by nearly a month between 1952 and 2000. Increased temperatures, reduced precipitation, and more frequent wildfires are likely throughout the Mediterranean. Precipitation has decreased over much of the region in recent decades, although the pattern of change has varied greatly from place to place.

In southern Canada, mean annual temperature increased by 0.5–1.5°C during the 1900s, with greatest warming in spring. In the northeastern United States, mean annual temperature has increased 0.25°C per decade since 1970. Less precipitation is falling as snow, lakes are thawing 9–16 days earlier in spring, and river flows peak earlier in spring. Climate models project additional warming of 2.1–2.9°C by midcentury, with summer warming somewhat greater than that in winter. Winters are projected to be wetter, but little or no increase in summer rainfall appears likely. The growing season is likely to be extended by a month or more, but the frequency of droughts is also likely to increase.

In western North America, models suggest that during this century, California will warm substantially, more in summer than winter. Precipitation is expected to decrease, especially in the Central Valley and northern coastal region. The winter snowpack in the Sierra Nevada will likely be reduced and subject to earlier melting, increasing the risk of wildfire and reducing freshwater inflow to areas such as the Sacramento–San Joaquin Delta.

Far to the south, tropical rain forest regions worldwide have warmed by about 0.26°C per decade since 1976. In the Amazonian region of South America, where the climate has warmed at the above rate, a total warming of 3.3°C is likely during this century. Since the early 1970s, rainfall has declined and the dry season has intensified in northeastern Amazonia, probably related to the increasing frequency and intensity of El Niños. Southeastern Amazonia and the Guyanas are also likely to see drier conditions during this century. In drier forests, fire may become increasingly important and further degrade these communities into low-biomass, fire-dominated forests. Many of these regions are experiencing rapid deforestation, which may increase the probability of drier conditions. Deforestation in the Amazon Basin appears likely to lead to longer dry seasons and reduced precipitation almost everywhere. Some scientists suggest that the interaction of climate change with deforestation will eventually lead to complete loss of these tropical forests. Such a change could increase warming to as much as 8°C. In any case, wintering migratory birds from both North and South Temperate regions, as well as intratropical migrants, will be severely affected by changes in Amazonia.

Climatic warming is somewhat less in southeastern Asia than in other rain forest areas, averaging 0.22°C per decade. This region has seen a moderate decline in precipitation, averaging about 1.0 percent per decade. Tropical Africa has warmed at a rate averaging 0.26°C per decade since 1976. African rain forests have also experienced more severe drying than other parts of the tropics. Precipitation has declined by about 2.4 percent per decade, with the decline being higher in West Africa and the northern Congo Basin than in the southern Congo Basin.

In the Sahelian region of Africa, subject to intense drought in the 1950s through 1980s, a significant increase in rainfall occurred in the 1990s. Future trends in this region are uncertain, but some models suggest that the warming of the North Atlantic Ocean may lead to stronger monsoons in the region through the mid-twentieth century.

Warming of the Southern Hemisphere is somewhat less than that of the Northern Hemisphere. Estimates in the late 1990s suggested a warming of 0.5–0.6°C for land regions, but more recent studies suggest greater warming. Australia has seen an increase of about 1°C in annual mean temperature since 1910, with the greatest increase being since 1980. Nighttime temperatures have increased most, and the duration of freezing temperatures has declined. Since 1950, rainfall has tended to increase in the northwest and decline in the southeast. In South America, Argentina has warmed about 1.0°C over the past century, with this change spread throughout the year. In southern Africa, global climate models predict an increase of 2–4°C with a doubling of atmospheric $CO_2$. In South Africa itself, a doubling of atmospheric $CO_2$ is projected to lead to general warming of 2.5–3.0°C. Warming may be greater in the northern interior region and less in coastal areas. How future precipitation is likely to change in South Africa is uncertain.

Pacific islands have already experienced substantial climatic change. On Oahu, Hawaii, for example, the mean temperature has increased 2.4°C and precipitation has decreased about 20 percent over the last 90–100 years. Projections of climate models suggest that temperatures in Hawaii could increase an additional 1.7°C, although how precipitation would change is uncertain.

Climatic change has dramatically affected Antarctica. The Antarctic Peninsula warmed about 2.6–3.0°C between 1945 and 1997. During midwinter the change along the peninsula was even greater, roughly 4–5°C. As we shall see, populations of several penguins and other seabirds on the Antarctic Peninsula are being differentially affected by this warming. Much of West Antarctica has also warmed, but parts of East Antarctica apparently have experienced significant cooling.

## Climate Change and Freshwater Ecosystems

In temperate and Arctic North America, climatic warming is already causing significant changes in the dynamics of lakes, wetlands, and streams utilized by numerous migratory water birds. In the Experimental Lakes

Area in northwestern Ontario, Canada, the regional mean air temperature warmed by 1.6°C between 1970 and 1990. Water temperatures have increased in all lakes to varying degrees, and they now experience a longer ice-free season, largely due to earlier spring disappearance of ice. Warmer temperatures and the longer ice-free season have increased evaporation and transpiration losses by about 90 millimeters over the 20-year period.

The dynamics and chemistry of northern lakes are also being altered. Changes in water temperature and ice cover modify patterns of lake stratification and overturn. In some parts of the northern United States, lakes that formerly exhibited a short period of ice cover are tending to become completely ice free, so lake overturn lasts throughout the winter. Farther north, lakes that normally do not stratify may undergo stratification in summer. Warmer conditions over the watersheds of lakes are reducing the inflow of water and nutrients and likewise reducing outflows. As a result, lake chemistry is changing, with an accumulation of ions, such as sodium and chloride, and a shift toward greater alkalinity.

Mountain areas in both the Northern and the Southern Hemisphere are experiencing climatic change. Most mountain glaciers in western North America lost 10–50 percent of their ice over the final three decades of the twentieth century. Water from these glaciers feeds mountain streams, so their shrinkage or loss will lead to reduced stream flows. Winter snowpack in the mountains of western North America has also diminished. Earlier warming of mountain areas has also led to earlier snowmelt, so peak streamflows are now tending to occur earlier in the year. Mountain regions of northern South America have warmed about 0.33°C per decade since the mid-1970s, and glaciers in Colombia, Ecuador, and Peru are retreating rapidly.

## Climate Change and the Marine Environment

Marine environments worldwide are being affected by climatic change, with results that interact with climate change on land.

### Sea Level Rise, Ice Cap Melting, and Sea Ice Retreat

Ocean levels are rising. Between AD 1900 and 2000, sea level rose an estimated 10 to 20 centimeters. Thermal expansion accounted for about half of this rise, with the rest due to increased inflow of water from the melting of ice on land. From the early 1950s through early 1990s, sea level rose an average of about 1.8 centimeters per decade. Recent satellite studies indicate that the rate has increased, so between 1993 and 2003 sea level rose 3.1 centimeters. As we noted earlier, substantial future sea level rise is certain.

Much of the recent rise in sea level appears to be due to an increase in meltwater and glacial discharge from the Greenland and Antarctic ice sheets. Between 2002 and 2009, losses from the Greenland ice sheet increased from 137 to 286 billion tons per year, and from the Antarctic ice sheet, from 104 to 246 billion tons per year. In 2005, net loss of Greenland ice alone accounted for 0.57 millimeters of annual sea level rise. The accelerating discharges from Greenland and Antarctica imply an annual acceleration of sea level rise of about 0.17 millimeters. Thus, earlier studies may have substantially underestimated the contribution of melting ice from Greenland to sea level rise.

The Antarctic ice sheet is also contributing to sea level rise, particularly through melting and glacial discharge in West Antarctica and the Antarctic Peninsula. The recent breakup of ice shelves along the Antarctic Peninsula, beginning in 1995, has dramatically increased discharge rates of neighboring coastal glaciers. This breakup continued into 2008, with the disintegration of a large area of the Wilkins Ice Shelf along the western Antarctic Peninsula. These inputs of water to the oceans are partially offset, however, by increased snow accumulation in the East Antarctic Ice Sheet. Of long-term concern is the possibility of collapse of a substantial portion of the West Antarctic Ice Sheet, which would likely lead to sea level rise of 3 or more meters.

In the Arctic Ocean the ice pack has decreased in area and thickness. Since 1978, the average area of sea ice has declined at about 2.7 percent per decade, with the area in summer declining 7.4 percent per decade. During 2007–2008 the ice pack reached its smallest recorded area. Thinning and earlier retreat of the ice pack is leading to major changes in pelagic and coastal ecosystems, as well as in seabird distributions. Coastal areas of Canada and Russia have also experienced summer ice-free conditions since the late 1970s. Freshwater inflows to the Arctic Ocean have increased, and ocean currents into and out of the Arctic Basin have changed.

## Ocean Warming

The world oceans, from the surface to a depth of 300 meters, have warmed by 0.31°C since the middle of the last century. Warming is dramatic in surface waters of the North Sea, between the British Isles and mainland Europe, where water temperature has increased 0.6°C since the early 1960s. Deep waters have warmed in some places, such as the Labrador Sea and adjacent North Atlantic Ocean, where warming is greater than in the surface 300 meters. In the southern ocean, waters of the Antarctic Circumpolar Current have also warmed about 0.17°C at depths of 700 to 1100 meters since 1950. This warming exceeds that of deep waters of other ocean regions. Ocean warming has affected invertebrates and fish on which seabirds depend, and populations of a number of migratory seabirds have declined alarmingly as a result.

Warming of tropical and temperate oceans is also associated with an increased frequency of high-intensity hurricanes, those of categories 4 and 5. This pattern, detected through analyses of hurricanes from 1970 to 2004, is evident in all of the major tropical and temperate oceans.

### Changes in Oceanic Current Systems

Climatic warming may be causing substantial changes in ocean current patterns. Since the 1970s, for example, warm conditions in the California Current, typical of the positive phase of the Pacific Decadal Oscillation, have intensified. This is evident in the greater abundance of tropical species, ranging from phytoplankton to fish and seabirds, in this oceanic region. Changes in land–sea temperature relations are apparently modifying upwelling patterns in several ocean areas, including the Arabian Sea and along the northwest African coast.

Warming of deep waters of the Antarctic Circumpolar Current is also associated with a southward shift of the current axis by about 50 kilometers in the Atlantic, Indian, and Pacific Ocean fronts.

### Ocean Acidification

About a quarter of anthropogenic $CO_2$ now enters the oceans, leading to depletion of carbonate ions and acidification of the water. As a result, the pH of the oceans has dropped about 0.1 pH unit over the past few decades. If $CO_2$ emissions continue as projected, a further drop of 0.3–0.5 units appears likely by AD 2100. Moreover, if atmospheric $CO_2$ levels exceed 500 ppm, major damage to coral reefs and their associated biotas is likely. Some reefs may already be suffering. Coral calcification on the Australian Great

Barrier Reef has declined about 14.2 percent since 1990. Although the exact cause is not fully understood, high temperature stress and decreased aragonite saturation state, reflecting increased acidity of seawater, are the likely major contributors. In the Arctic Ocean, acidification and melting sea ice are reducing carbonate ion availability for plankton, shellfish, and finfish.

## Summary

Increase of greenhouse gases in the atmosphere is leading to climate changes in all global realms: continental areas, freshwaters, and oceans. Global warming is now significant at all latitudes, and even at current levels of greenhouse emissions, the commitment for future warming is considerable. The most profound climate changes are occurring at high latitudes and are greatest in the Northern Hemisphere, where many bird species show some pattern of seasonal migration. Nevertheless, even in the tropics, substantial changes are occurring. Major systems of atmospheric and oceanic circulation are likely being altered by global warming. Freshwater ecosystems are also experiencing warming, which is altering patterns of ice cover and seasonal change. Marine ecosystems are likewise experiencing changes in winter ice cover, as well as in current systems, with effects on many migratory seabirds. Acidification of ocean water will likely have severe effects on corals and other marine life.

### KEY REFERENCES

De'ath, G., J. M. Lough, and K. E. Fabricius. 2009. "Declining coral calcification on the Great Barrier Reef." *Science* 323:116–119.

Delbart, N., G. Picard, T. Le Toans, L. Kergoats, S. Quegan, I. Woodward, D. Dye, and V. Fedotova. 2008. "Spring phenology in boreal Eurasia over a nearly century time scale." *Global Change Biology* 14:603–614.

Canadell, J. G., C. Le Quere, M. R. Raupach, C. B. Field, E. T. Buitenhuis, P. Ciais, T. J. Conway, N. P. Gillett, R. A. Houghton, and G. Marland. 2007. "Contributions to accelerating atmospheric $CO_2$ growth from economic activity, carbon intensity, and efficiency of natural sinks." *Proceedings of the National Academy of Sciences USA* 104:18866–18870.

Cook, A. J., A. J. Fox, D. G. Vaughan, and J. G. Ferrigno. 2005. "Retreating glacier fronts on the Antarctic Peninsula over the past half-century." *Science* 308:541–544.

Gille, S. T. 2002. "Warming of the Southern Ocean since the 1950s." *Science* 295:1275–1277.

Hayhoe, K., C. P. Wake, T. G. Huntington, L. Luo, M. D. Schwartz, J. Sheffield, E. Wood, et al. 2006. "Past and future changes in climate and hydrological indicators in the U.S. Northeast." *Climate Dynamics* 28:381–407.

Mahli, Y., J. T. Roberts, R. A. Betts, T. J. Killeen, W. Li, and C. A. Nobre. 2008. "Climate change, deforestation, and the fate of the Amazon." *Science* 319:169–172.

Osborn, T. J. and K. R. Briffa. 2006. "The spatial extent of 20th-century warmth in the context of the past 1200 years." *Science* 311:841–844.

Raupach M. R., G. Marland, P. Ciais, C. Le Quere, J. G. Candekk, G. Klepper, and C. B. Field. 2007. "Global and regional drivers of accelerating $CO_2$ emissions." *Proceedings of the National Academy of Sciences USA* 104:10288–10293.

Schindler, D. W. 1997. "Widespread effects of climatic warming on freshwater ecosystems of North America." *Hydrological Processes* 11:1043–1067.

Solomon, S., G.-K. Plattner, R. Knutti, and P. Friedlingstein. 2009. "Irreversible climate change due to carbon dioxide emissions." *Proceedings of the National Academy of Sciences USA* 106:1704–1709.

Steig, E. J., D. P. Schneider, S. D. Rutherford, M. E. Mann, J. C. Comiso, and D. T. Schindel. 2009. "Warming of the Antarctic ice-sheet surface since the 1957 International Geophysical Year." *Nature* 457:459–462.

Stenseth, N. C., G. Ottersen, J. W. Hurrell, A. Mysterud, M. Lima, K.-S. Chan, N. G. Yoccoz, and B. Ådlandsvik. 2003. "Studying climate effects on ecology through the use of climate indices: The North Atlantic Oscillation, El Niño Southern Oscillation and beyond." *Proceedings of the Royal Society of London B* 270:2087–2096.

Velicogna, I. 2009. "Increasing rates of ice mass loss from the Greenland and Antarctic ice sheets revealed by GRACE." *Geophysical Research Letters* 36:L19503, doi:10.1029/2009GL040222.

Wigley, T. M. L. 2005. "The climate change commitment." *Science* 307:1766–1769.

# Chapter 3

# *Global Climate Change and Alteration of Migratory Bird Habitats*

In 1913, as the Panama Canal neared completion, a dam on the Chagres River impounded Gatun Lake, at the time the largest artificial lake in the world. Formation of the lake also created Barro Colorado Island from an area of upland tropical forest 1564 hectares in size. Since 1923, the Smithsonian Institution has managed the island for tropical research. I carried out my doctoral research on the island in 1958 and 1959 and returned in 1970 for studies of tropical birds and wintering North American migrants. Although I realized then that the ecology of the island was changing, I did not recognize the role of changing climate. Others soon did, however, and in 1980, Smithsonian and Princeton University scientists established a 50-hectare Forest Dynamics Plot near the center of the island. As we shall see, periodic surveys on this plot are now revealing that this tropical forest is indeed being affected by global climatic change.

Changing global climate is altering the habitats of migratory birds at a rate faster, in all likelihood, than at any previous time in earth's history. For many, if not most, migratory birds, climate change is differentially affecting areas that are used during breeding and nonbreeding seasons and in migration. Here, we will examine these effects so that we can consider the ability of migratory birds to respond adaptively. We will then ask whether most migratory birds are at greater risk of population decline or extinction than resident species because they depend on use of specific habitats at different seasons. On the other hand, we will consider whether their ability to use different habitats and resources at different seasons enables them to adapt to changing conditions more easily than resident species.

In this chapter we shall consider how climatic change is affecting plant communities and biological productivity in land, freshwater, coastal and estuarine, and marine environments. In the next chapter we shall examine other major types of environmental change caused by direct human influences on the environment.

## Terrestrial Environments

All terrestrial ecosystems are being modified by climatic change, with implications for migratory birds throughout their annual cycles.

### The Tropics

Tropical forests are experiencing substantial greenhouse warming. How these ecosystems respond is critical, since they are wintering habitats for many migratory birds from higher latitudes, as well as the home of many intratropical migrants. The overall effects of this warming are not yet fully understood, although it is clear that the structure and dynamics of forests are being altered. Between 1982 and 1999, ground and satellite-based observations seemed to indicate that net primary production in tropical regions increased about 7.4 percent, although it showed strong year-to-year fluctuations related largely to El Niño events. Recruitment and mortality of trees are also reported to be increasing throughout the tropics, leading to a higher rate of turnover of tree populations.

Although the overall rate of net primary production in tropical forests does not appear to differ from that of temperate forests, the rate of increase in primary production in most areas of the tropics appears to exceed that at higher latitudes. This increase is consistent with a fertilizing effect of increased atmospheric $CO_2$. Across the tropics, recent analyses suggest that remaining tropical forests realize a net storage of 1.3 billion tons of carbon annually (Table 3.1). This represents a substantial sink for anthropogenic $CO_2$.

#### NEOTROPICS

The climate of the Neotropical forest region has warmed, but change in precipitation has been variable from place to place. Considerable uncer-

TABLE 3.1. Annual increase in biomass of tropical forests during 1987–1997.

| Region | Area ($10^6$ ha) | Annual Tree Carbon Storage ($10^9$ tons) | Annual Tree Carbon Storage ha$^{-1}$ (tons) |
|---|---|---|---|
| Americas | 786.8 | 0.62 | 0.79 |
| Africa | 632.3 | 0.44 | 0.70 |
| Asia | 358.3 | 0.25 | 0.70 |
| Total | 1777.3 | 1.31 | 0.74 |

Source: Data from S. L. Lewis et al. 2009.

tainty exists about the effects of warming on forest dynamics, however, with analyses based on satellite data and those on ground-level measurements often reaching differing conclusions. The complex history of natural and human disturbance of tropical forests also makes interpretation of short-term changes in measures of growth risky.

Both satellite and ground-based studies have attempted to measure net primary production of mature forests. Some studies suggest that climatic change has increased the productivity of these forests substantially. Between 1982 and 1999, for example, satellite data combined with ground observations suggested that net primary production in the Amazon Basin had increased over 1 percent annually, with this region accounting for 42 percent of the total world increase in net primary production. This conclusion, however, is subject to uncertainties about accuracy of these techniques of measurement.

Ground-based studies have included short-term measurements of gas exchange in forest plots and long-term measurements of tree growth. Gas exchange studies have yielded conflicting results, perhaps reflecting the technical difficulties of measuring gas fluxes in large, open systems. In the Amazon Basin of South America, tree biomass was measured in a network of mature forest plots beginning in the 1980s. Data from these plots show that aboveground biomass increased through 2004, as expected from a fertilizing effect of increasing $CO_2$. Both tree growth rates and mortality rates appeared to have increased, with biomass increase due to growth exceeding that lost due to mortality. Recruitment of young saplings also appeared to exceed mortality, so both tree density and total forest biomass seemed to be increasing. In addition, the abundance and biomass of lianas increased by nearly 40 percent in undisturbed forest stands between 1981 and 2001, changing an important feature of forest structure. This increase may be due to a greater growth response of lianas to increased $CO_2$ concentration.

On the other hand, direct measurements in old-growth rain forests in Central America give a different picture. At La Selva, Costa Rica, from 1984 through 2000, measurements on six important tree species indicated that growth of trees had declined. Growth rate decreased in linear fashion as daily minimum temperatures increased for all species. In addition, during the very warm El Niño years of 1987–1988 and 1997–1998, tree growth appeared to be greatly depressed. In another series of eighteen plots at La Selva, on which more detailed biomass measurements were made from 1997 to 2001, some plots gained and some lost biomass, with no clear trend being apparent over this short interval.

Similar measurements on the Forest Dynamics Plot at Barro Colorado Island, Panama, from 1985 through 2005 also suggest that tree growth rates declined. Declines in growth were noted for the majority of tree species and for all size classes. These declines were correlated with an increase in daily minimum temperatures and an increase in regional cloudiness.

Results of these Central American studies thus conflict with those from Amazonia. Neotropical forests are almost certainly being modified by changing climate, however, and the important wintering areas for migratory birds in Central America may be showing declines in growth.

Influences of warming tropical oceans may also exert major influences on Neotropical forest ecology. In 2005, the Amazon Basin experienced severe drought due to greatly reduced rainfall and warmer than normal temperatures. Conditions were driest and warmest in southern and western Amazonia. In addition to reduced water levels in the major rivers, many floodplain lagoons dried out, and extensive forest fires occurred. Warmer waters in the tropical Atlantic may have caused this drought and may also have been the underlying cause of the severe hurricane season of 2005. The drought in 2005 led to a sudden, large decline in biomass in forest plots throughout the Amazon Basin.

The dramatic drought of 2005 may be a forerunner of changes in Amazonia, as we noted in the last chapter. Climate models predict drier conditions in parts of the Amazon Basin that, at the very least, may lead to conversion of rain forests to seasonal forests during the coming century. Changes of this magnitude would affect many intratropical migrants and wintering temperate zone migrants.

### SUB-SAHARAN AFRICA

In tropical Africa, in spite of warming and drying of the rain forest region, analysis of aboveground carbon storage in forest plots in ten countries shows that substantial increases in biomass occurred between 1968 and 2007. Overall, this analysis suggests that each hectare of forest trapped about 0.7 tons of carbon per year. Maturation of these forests may benefit some intratropical forest migrants, although some wintering species from Europe may be affected adversely by the drier conditions.

## ASIAN–AUSTRALIAN TROPICS

Southeast Asian rain forests, which have also warmed, seem to be accumulating biomass. Here, however, a strong relation exists with the El Niño events that lead to reduced precipitation, especially in the region from Borneo to New Guinea. The increasing frequency and severity of these droughts has led to severe impacts on forest vegetation. In 1997–1998, for example, the very severe El Niño led to extensive tree mortality, as well as widespread wildfires. Pollinator wasps of eight species of fig trees also became locally extinct because of failure of fruit production and were slow to recolonize the trees. The reduction in availability of key foods for many frugivorous birds and other animals is an example of how the changing El Niño pattern can alter forest ecology in this region.

The tropical forests of northern Queensland, Australia, are less severely affected but will likely experience major changes by the middle of the twenty-first century. Queensland exhibits a mosaic of forest types, ranging from dry woodlands to humid lowland rain forest and highland cloud forest. Modeling studies suggests that change in temperature and rainfall patterns will cause substantial shifts in the area and spatial distribution of lower-elevation forest types. Some montane birds that move to lower elevations in the nonbreeding season, such as the Noisy Pitta, might be affected by such forest shifts.

## TROPICAL HIGHLANDS

Global climate models suggest that warming associated with a doubling of atmospheric $CO_2$ will affect cloud forest areas worldwide. During the dry season, humidity levels essential for cloud forest survival will likely be shifted upward by 300 meters or more in Central and South America, Borneo, and Africa. In Queensland, Australia, with a mean climatic warming of 1.0°C, wet highland forests are likely to decrease in area by 50 percent.

At high altitudes in the Neotropics, cloud forests in some areas have already experienced warmer and drier conditions. The Monteverde Cloud Forest Reserve in the Tilarán Mountains of Costa Rica, for example, has apparently responded to warming sea surface temperatures by a substantial drying. The warming ocean has increased evaporation in the mountains. As moist air is carried up in elevation, orographic condensation releases more heat, warming the higher elevations of the mountains. This, in turn, leads to an increase in elevation of the cloud base, reducing the frequency of dense cloud mist that often permeates the high-elevation forests. During the dry season, especially, this leads to drier conditions. At Monteverde, the number of dry days has increased more than threefold since the early 1970s, and streamflow has decreased by more than 50 percent. Although climatic warming could potentially favor increased productivity in these montane forests, reduction in moisture availability is likely to offset such an increase. Many species of altitudinal migrants are already being affected by these climatic changes.

## Mid and High Latitudes of the Northern Hemisphere

Warming of the Northern Hemisphere appears to have advanced the spring season and increased plant growth in high latitudes. Satellite radiometers also show earlier greening of landscapes in Arctic, subarctic, and North Temperate regions around the globe. Plant and animal phenologies have also responded to lengthened seasons in many areas.

### NORTH AMERICA

In the North American Arctic, the growing season is now longer. In Canada and Alaska, satellite radiometer data show that plant growth has increased all across the continent. Climatic warming is leading to rapid vegetational change in some areas. Shrub cover has increased in tundra areas, and the conifer tree line is advancing into tundra in many areas, particularly the North Slope of Alaska. This change sets in motion a positive feedback relating to nutrient cycling and growth of woody plants. Shrub patches accumulate deeper layers of snow, which insulate the soil, slowing soil freezing and permitting microbial activity to continue longer into fall and winter. This process has increased the period of microbial activity by up to 2 months. In turn, shrubs are better able to utilize the nutrients, particularly nitrogen, made available by microbial activity, thus favoring their growth and spread. Forest and shrubland birds will clearly benefit from these changes but at the expense of species of the open tundra.

Farther south, forest vegetation has advanced into tundra in parts of Alaska and Canada. On the Alaskan Kenai Peninsula, drier and warmer conditions have encouraged the expansion of forest vegetation and an increase in tree line elevation. At the same time, many lakes, ponds, and muskeg areas have shrunk. In Quebec, Canada, the tree line has advanced about 12 kilometers on the eastern side of Hudson Bay.

Climatic warming is already significant in the deciduous forests of eastern North America, where January temperatures are predicted to increase 2.5–6.6°C and July temperatures, 2.3–5.0°C over the next century. These changes will shift the optimal geographic ranges of many plant species by 200 kilometers or more. How quickly plant species can track the changes in optimal range is uncertain. A modeling study of dispersal abilities of several tree species of the eastern United States suggests that colonization of areas 10–20 kilometers beyond the present range boundary is likely. Beyond this range, however, the probability of colonization becomes very small. How the numerous Neotropical migrants that breed in these forests will respond to lags in vegetation response is uncertain.

In western North America, mean annual temperatures have risen as much as 1–2°C since the late 1940s, with the greatest change being in the Pacific Northwest. Most areas have seen warmer winter and spring conditions and a lengthening of the frost-free season. Warming has led to earlier flowering by many plants, earlier peaks in spring snowmelt, and increased tree mortality in forests throughout the region. It has promoted altitudinal changes in plant activity in western mountains. In the Santa Catalina Mountains of southern Arizona, for example, over a quarter of the plant species showed altitudinal changes in their flowering ranges over a 20-year period, most shifting their flowering ranges upward. Migratory hummingbirds are likely to track such changes very closely. On the desert slope of the Santa Rosa Mountains of southern California, dominant plants have shifted their zone of greatest dominance upward, although not increasing their overall altitudinal distribution.

## EURASIA

The Mediterranean region is predicted to be the most vulnerable area of Europe to global climate change. Climatic warming and changes in rainfall patterns imply considerable modification of seasonal patterns of food availability for both short- and long-distance migratory birds.

Warming of high latitudes in Europe and Asia has had effects similar to those in North America. In central Sweden, the upper altitudinal limits of several species of trees have advanced by 120 to 375 meters. Spring soil thaw in Asian tundra has advanced several days per decade, and shrub expansion into tundra may also be occurring in a fashion similar to that in North America.

## Mid and High Latitudes of the Southern Hemisphere

Although clear evidence for climatic warming exists for all southern continents, documentation of biotic changes is scarce. In temperate South America, where mean annual temperature has increased by about 1.0°C, coastal and montane rain forests in Argentina and Chile are considered to be threatened by warmer and drier conditions. In Australia, climatic warming is affecting alpine areas of New South Wales, where winter snows have decreased substantially. Woody plants are also invading alpine and subalpine grasslands. In southern Africa, climatic warming and decreasing rainfall are predicted to push the region of winter rainfall southward. The arid Karoo region will be reduced in area, and the fynbos biome of the Cape Province area to the south will also be severely reduced in area.

In Antarctica, warming of the Antarctic Peninsula and reduction in ice and snow cover are already leading to an expansion of native vascular plants. These species, Antarctic Hairgrass (*Deschampsia antarctica*) and Antarctic Pearlwort (*Colobanthus quitensis*), occur on several islands along the western side of the peninsula.

## Oceanic Islands

Climatic warming will likely alter the extent and composition of Hawaii's unique montane forests. Hawaiian cloud forests are some of the wettest forest ecosystems on earth. Even small changes in climate could cause major changes in cloud cover and rainfall for these forests. Warmer temperatures would likely cause a shift in distribution of cloud forest species and could possibly increase the frequency of fire and alien plant invasion. The habitats of endemic honeycreepers, some of which are altitudinal migrants, may be altered by these influences. An increase in frequency or intensity of hurricanes will exacerbate these problems.

In tropical regions, one of the most serious threats to seabirds is survival of low-lying coral atolls in the face of rising sea levels. The Northwest

Hawaiian Islands form an 1800-kilometer-long chain of pinnacles, low islands, and coral atolls. Many of the bird nesting islands are less than 2 meters in maximum elevation. High-scenario projections of a sea level rise of
88 centimeters by AD 2100 suggest that the islets at French Frigate Shoals
and Pearl and Hermes Reef will likely lose 56 percent and 69 percent of
their area, respectively. Although, in principle, coral growth rates should be
able to keep pace with the projected sea level rise, changes in ocean chemistry and rising water temperatures may slow this response.

## Freshwater Environments.

Freshwater lakes, streams, and marshes—the breeding, migratory stopover,
and nonbreeding habitats for many waterbirds—are particularly vulnerable
to climatic change.

### Lakes and Streams

In the Arctic and subarctic, earlier and longer summers will cause freshwater systems to warm considerably. These waters will become fed more heavily by rain, as opposed to melting snow and ice. Rivers will likely experience
reduced influences of ice breakup and spring flooding. Warmer air and water temperatures will increase evaporation and transpiration and also lead to
increased productivity in lakes and streams. These changes are already being
shown by the world's largest freshwater lake, Baikal in Siberia, which is experiencing warming and increased primary production.

   Climatic warming may cause the disappearance of many high-latitude
lakes and wetlands that owe their presence to underlying permafrost. Analysis of satellite images in Russian Siberia has shown that between 1973 and
2004, about 11 percent of 10,000 large lakes either disappeared completely
or shrank to very small areas. This pattern suggests that thawing of the permafrost has enabled rapid drainage into deep subsurface zones. Similar
changes are seen in parts of Alaska. The impact of such changes on migratory
waterfowl, shorebirds, and many other species would be substantial.

### Other Wetlands

In areas such as the Prairie Pothole Region of central North America,
warmer temperatures will likely lead to decreases in soil moisture, causing

the loss of many wetlands and shortening the seasonal duration of others. Annual variability in wetland occurrence and area will increase. The depth, salinity, temperature, aquatic plant structure, and composition of aquatic food webs will change. Waterfowl are thus likely to experience much-reduced breeding habitat. Modeled future conditions in the Prairie Pothole Region project significant declines in wetlands and in breeding duck populations by the 2080s.

## Coastal and Estuarine Habitats

Rising sea levels already are having impacts on coastal habitats used by many migratory birds. How biotic and physical conditions of beaches, mudflats, and salt marshes will respond is a complex issue. Coastal topography, sedimentation patterns, tidal influences, and marsh vegetation all influence the response to rising sea level. Nevertheless, in Jamaica Bay, New York, sea level rise is an important factor in the loss of marshland. In some situations, however, the fertilizing effect of $CO_2$ may stimulate vegetational growth and organic deposition that offset sea level rise. In many areas, change in coastal bird habitats will also be influenced by human efforts to defend low-lying land against erosion by rising seas.

## Oceanic Environments

Global warming is expected to lead to major changes in marine productivity. In tropical and midlatitude oceans, models suggest, productivity will decline, whereas at high latitudes, warmer waters, increased nutrient-rich river discharges, and stronger upwelling will enhance productivity.

### Arctic and Antarctic Regions

In the Arctic Ocean and Bering Sea, warming climate and reduction of sea ice cover are leading to major changes in the pelagic ecosystem. The phytoplankton community of the Arctic Ocean is changing, with smaller forms replacing larger species. This change will ultimately alter the pelagic food chain leading to marine birds. In the Bering Sea, the rich benthic invertebrate community of ice-dominated areas, which is exploited by marine mammals and birds, has declined where the ice pack has thinned and re-treated. Its place has been taken by a marine ecosystem dominated by pelagic fish.

In the Southern Hemisphere, with warming of sea and air tempera-tures around the Antarctic Peninsula, major changes have occurred in the marine food chain. We shall examine these changes in detail in Chapter 5.

## Midlatitude Oceans

In the northeast Atlantic, major changes in the pelagic ecosystem have oc-curred. Warm-water assemblages of plankton have expanded northward by about 1000 kilometers, with corresponding retreat of cold-water forms. Blooms of phytoplankton and zooplankton have also shifted their seasonal peaks in correspondence with water temperature conditions. In waters of the western North Atlantic fed by outflow from the Arctic Basin, phyto-plankton productivity has increased and the zooplankton community has become reorganized.

In the North Sea, these warmer waters have led to the northward shift of populations of a number of fish by distances ranging from 48 to 403 kilometers. Range shifts were particularly strong for small species with short life cycles. The impacts of these changes on seabirds are enormous, as we shall see.

In the Gulf of Alaska, a shift from a cold to a warm surface-water re-gime took place in 1977 and has continued to the present. This shift was correlated with a southward displacement of the Aleutian low-pressure cen-ter, which allowed westerly winds to push warm surface waters into the gulf. Under the altered regime, shrimp and Capelin (*Mallotus villosus*), on which many seabirds feed, have declined to very low levels.

Off the coast of California, the Pacific Decadal Oscillation has influ-enced surface waters of the California Current. From the late 1940s through the mid-1970s, the current experienced cool conditions and strong up-welling of nutrient-rich water. From 1978 through 1999, warm conditions with weak upwelling prevailed, and plankton abundance declined. In 2000, cool conditions returned, with strong upwelling in most years. In 2005, however, upwelling largely failed, and productivity of marine foods for seabirds declined catastrophically. Some models suggest that climatic warm-ing is likely to intensify the temperature gradient between land and ocean along the central Pacific coast, leading to stronger wind-induced upwelling.

## Tropical Oceans

In low-latitude, strongly stratified ocean areas, satellite observations suggest that ocean warming in the decade since about 1999 has led to declining primary productivity. Most ocean areas with waters that warmed more than 0.15°C between 1999 and 2004 showed a decline in primary production and stronger stratification. If global warming is indeed responsible for this trend, altered oceanic food chains in low-latitude seas may lead to major declines in upper-level predators, including seabirds.

Coral reefs and associated atolls are also at risk from the combined effects of ocean warming and acidification. Bleaching and mortality of reef-building corals is increasing at an alarming rate, especially in the East Indies and Caribbean. More than 30 percent of these species appear to be at an elevated risk of extinction. Substantial loss of these corals would alter the dynamics of islands used by nesting seabirds, compounding the impacts of rising sea level.

## Summary

Terrestrial environments at high latitudes are in many areas experiencing longer growing seasons and in some cases undergoing vegetational changes involving spread of woody plants. Elsewhere, drier conditions and increased severity of storms may impact bird habitats. Freshwater ecosystems are likely also to experience longer warm seasons, but they may also shrink in size or disappear. Estuarine areas and oceanic islands may be affected by rising sea levels, and ocean areas themselves are likely to experience altered patterns of currents and productivity. All of these changes will affect the habitats of migratory birds.

### KEY REFERENCES

Anderson, P. J. and J. F. Piatt. 1999. "Community reorganization in the Gulf of Alaska following ocean climate regime shift." *Marine Ecology Progress Series* 189:117–123.

Atkinson, A., V. Siegel, E. Pakhomo, and P. Rothery. 2004. "Long-term decline in krill stock and increase in salps within the Southern Ocean." *Nature* 432:100–103.

Behrenfeld, M. J., R. T. O'Malley, D. A. Siegel, C. R. McClain, J. L. Sarmiento, G. C. Feldman, A. J. Milligan, P. G. Falkowski, R. M. Letelier, and E. S. Boss.

2006. "Climate-driven trends in contemporary ocean productivity." *Nature* 444:752–755.

Carpenter, K. E. and 32 others. 2008. "One-third of reef-building corals face elevated extinction risk from climate change and local impacts." *Science* 321:560–563.

Feeley, K. J., S. J. Wright, M. N. N. Supardi, A. R. Kassim, and S. J. Davies. 2007. "Decelerating growth in tropical forest trees." *Ecology Letters* 10:461–469.

Grebmeier, J. M., J. E. Overland, S. E. Moore, E. V. Farley, E. C. Carmack, L. W. Cooper, K. E. Frey, J. H. Helle, F. A. McLaughlin, and S. L. McNutt. 2006. "A major ecosystem shift in the northern Bering Sea." *Science* 311:1461–1464.

Hampton, S. E., L. R. Izmest'eva, M. V. Moore, S. L. Katz, B. Dennis, and E. A. Silow. 2008. "Sixty years of environmental change in the world's largest freshwater lake: Lake Baikal, Siberia." *Global Change Biology* 14:1947–1958.

Huntingford, C., R. A. Fisher, L. Mercado, B. B. B. Booth, S. Sitch, P. P. Harris, P. M. Cox, et al. 2008. "Towards quantifying uncertainty in predictions of Amazon 'dieback'." *Philosophical Transactions of the Royal Society B* 363:1857–1864.

Huston, M. A. and S. Wolverton. 2009. "The global distribution of net primary production: Resolving the paradox." *Ecological Monographs* 79:343–377.

Lewis, S. L. and 32 others. 2009. "Increasing carbon storage in intact African tropical forests." *Nature* 457:1003–1006.

Phillips, O. L., S. L. Lewis, T. R. Baker, K.-J. Chao, and N. Higuchi. 2008. "The changing Amazon forest." *Philosophical Transactions of the Royal Society B* 363:1819–1827.

Sturm, M., J. Schimel, G. Michaelson, J. M. Welker, S. F. Oberbauer, G. E. Liston, J. Fahnestock, and V. E. Romanovsky. 2005. "Winter biological processes could help convert arctic tundra to shrubland." *BioScience* 55:17–26.

# Chapter 4

# *Other Global Change Threats to Migratory Bird Habitats*

When I first visited Costa Rica in 1961 as a participant in a summer program in tropical biology, natural forests covered somewhere between 55 and 60 percent of the country. The human population then was about 1.3 million. Flying from San Jose east to the Valle del General, our group saw only unbroken forest. Farther on, near the Panamanian border, however, we visited a large coffee plantation that was being developed on the slopes of the Talamanca Range. Most of the forest was being cut, with some large trees being left for shade. The understory was removed to make way for coffee plants. Since then, Costa Rica's population has grown to over 4.4 million, and extensive plantations have been developed for coffee, bananas, oil palm, timber trees, pineapple, cacao, blackberries, and many other tropical crops. Total forest area is now reported to be slightly over 46 percent, but this includes large areas of second growth, coffee farms, and tree plantations. Costa Rica has developed a remarkable network of natural preserves, but the tropical forests, home to many intratropical migrants and wintering temperate zone species, are greatly altered in most places.

Human activities are now important worldwide, leading to disturbance and fragmentation of natural vegetation types by harvesting, clearing, cultivation, and grazing. Fragmentation makes small units of habitat subject to outside physical and biotic influences. Overgrazing and other intensive uses can lead to desertification in many semiarid regions. Many land use practices can modify natural wildfire regimes. Alien species, both birds and other species, have increasingly become major disruptive factors for native communities. Pollution by nutrient-rich wastes and synthetic bioactive chemicals is still another influence of global extent.

These impacts on natural ecosystems interact with climatic processes to compound the influences of global change on migratory birds. They produce both physical feedbacks, by altering energy exchange between the atmosphere and land and water ecosystems, and chemical feedbacks, by changing the rates of entry of greenhouse gases and other chemicals into the biosphere. Some activities also alter the intensity of bioactive components of solar radiation, such as those in the ultraviolet range.

## Destruction and Fragmentation of Natural Ecosystems

Agricultural and urban developments have transformed about 29 percent of the earth's natural ecosystems, particularly grasslands, savannas, and forests. About a quarter of the world's native grasslands have become croplands or managed pastures. Savanna communities are suffering rapid conversion, especially in Africa. Very little virgin forest survives, especially in the temperate zone. Little change has actually occurred in the total area of forest in the temperate zone and subarctic. Only about 3 percent of temperate forest, however, can be considered natural and relatively undisturbed. For boreal forests, in contrast, about 48 percent of forest area is probably relatively undisturbed. Although large areas of tropical forests still exist, only about 44 percent can be considered relatively undisturbed.

In addition, worldwide, forest communities are fragmented into stands of varying size and are surrounded by nonforest vegetation or landscapes of human creation. Weather conditions at the edges of these stands differ from those of the stand interiors. Edge zones are subject to invasion by nonforest species and the biotic influences of species resident in the surrounding habitat. For many migratory birds, these influences translate into increased brood parasitism and nest predation, leading to lower reproductive success in small forest patches. Some small subpopulations may never achieve a reproduction adequate to offset mortality and thus are "sink" populations that can only be maintained by dispersal from larger populations that have a positive population growth rate.

Habitat fragmentation also alters the dynamics of populations of birds and other organisms because of the many random factors that act on populations. In subpopulations that are small and more isolated, extinction and colonization become dominant processes. The smaller the habitat patch, the greater is the probability of local extinction, and the greater the isolation, the lower is the probability of recolonization across gaps.

Analyses with satellite imagery reveal the extent of fragmentation of forest communities. These analyses characterize vegetation as "forest" or "nonforest" on pixels 1 square kilometer in area. The overall degree of fragmentation is estimated based on the number and pattern of forest and nonforest pixels in units of 81 square kilometers. Forest units are characterized as interior if all pixels are forest, as edge if a boundary with nonforest exists, or as an island if forest is surrounded by nonforest. Fragmented units are partitioned into those due to natural causes and those due to human clearing of forest, by considering whether forest edges separate forest from another natural vegetation type or from areas of human land use.

Although these analyses are at a coarser scale than some previous analyses of forest fragmentation, they reveal a number of differences relevant to migratory bird species. For the world as a whole, only about one-third of forest qualifies as interior, because of either natural habitat conditions or human alteration. Among major continental regions (Table 4.1), North America has the highest percentage of interior forest and the lowest percentage of forest islands. Africa and the Australia–Pacific regions show low percentages of interior forest and high percentages of forest islands.

Habitat fragmentation is a severe problem in all major communities. Fragmentation due to human activities affects more than half of temperate broadleaf and Mediterranean forests and nearly half of dry broadleaf forests in the tropics (Table 4.2). Fragmentation due to human activities exceeds that from natural causes for all major forest types except boreal forests and taiga.

TABLE 4.1. Forest fragmentation patterns for major world regions.

| Region | Area ($10^3$ km²) | Interior Forest (%) | Forest with Edge (%) | Forest with Openings (%) | Forest Islands (%) |
|---|---|---|---|---|---|
| North America | 8565 | 44.9 | 14.0 | 28.5 | 4.7 |
| South America | 6940 | 33.0 | 11.6 | 39.6 | 6.5 |
| Eurasia | 9551 | 32.0 | 14.1 | 34.1 | 8.7 |
| Africa | 2732 | 28.7 | 13.9 | 31.8 | 12.7 |
| Australia/Pacific | 2135 | 27.1 | 16.1 | 35.1 | 8.7 |

Source: Modified from Riitters, K., J. Wickham, R. O'Neill, B. Jones, and E. Smith. 2000. "Global-scale patterns of forest fragmentation." Conservation Ecology 4(2):3. http://www.consecol.org/vol4/iss2/art3/. Transitional and undetermined categories are excluded.

TABLE 4.2. Average percentages of forest fragmentation due to anthropogenic ($F_a$) and natural ($F_n$) causes. Higher values indicate greater fragmentation, and the ratio $F_a/F_n$ indicates the importance of anthropogenic fragmentation relative to natural fragmentation.

| World Region | Africa | | Asia | | Australia | | Europe | | North America | | South America | | World | |
|---|---|---|---|---|---|---|---|---|---|---|---|---|---|---|
| Plant Community | $F_a$ | $F_n$ | $F_a$ | $F_n$ | $F_a$ | $F_n$ | $F_a$ | $F_n$ | $F_a$ | $F_n$ | $F_a$ | $F_n$ | $F_a$ | $F_n$ |
| Tropical/Subtropical | | | | | | | | | | | | | | |
| Moist Broadleaf | 18.3 | 8.6 | 48.4 | 3.7 | 29.8 | 6.3 | | | 30.8 | 6.4 | 13.4 | 4.2 | 24.8 | 5.2 |
| Dry Broadleaf | 24.2 | 35.7 | 75.7 | 1.0 | 44.3 | 1.6 | | | 31.1 | 5.1 | 24.7 | 11.4 | 48.8 | 6.5 |
| Coniferous | | | 44.5 | 13.7 | 66.8 | 0.2 | | | 17.9 | 8.4 | | | 21.7 | 8.9 |
| Mangrove | 26.2 | 29.4 | 74.0 | 4.2 | 22.4 | 4.8 | | | 41.1 | 13.6 | 30.0 | 7.9 | 39.3 | 12.5 |
| Temperate | | | | | | | | | | | | | | |
| Broadleaf/Mixed | 19.4 | 24.6 | 49.9 | 8.5 | 17.8 | 15.0 | 73.7 | 1.5 | 33.7 | 0.2 | 7.0 | 18.0 | 52.8 | 4.4 |
| Coniferous | 30.8 | 20.5 | 13.3 | 24.0 | | | 45.0 | 9.6 | 11.6 | 9.4 | | | 15.9 | 13.6 |
| Mediterranean | 35.4 | 36.0 | | | 45.6 | 22.1 | 68.8 | 9.6 | 6.3 | 26.7 | 66.0 | 8.3 | 55.4 | 16.9 |
| Boreal | | | | | | | | | | | | | | |
| Forest/Taiga | | | 3.0 | 11.5 | | | 12.9 | 4.4 | 1.2 | 21.6 | | | 4.2 | 13.1 |

Source: Modified from Wade et al. 2003.

## North America and Eurasia

Most vegetation types throughout North America and Eurasia show substantial fragmentation. With changing climatic conditions at higher latitudes, fragmentation becomes a serious impediment to range shifts by plants in the face of climatic change. Range expansions into new climatically favorable areas thus may not offset extinctions of populations in areas that have become unfavorable. The habitats of migratory birds may thus be slow to adjust plant structure to changing climate.

### BREEDING HABITATS

For continental North America (Table 4.2), which includes Mexico and Central America, fragmentation varies considerably for different vegetation types. For the various conifer forests of Canada, Alaska, and the western United States, total forest fragmentation averages about 21–23 percent. Temperate broadleaf and mixed forests show higher percentages of fragmentation. Tropical forests in general show high levels of fragmentation. Thus, fragmentation is clearly a severe problem in north temperate forests, where many migratory birds breed, and in tropical forests, where many winter.

In deciduous forests of eastern and midwestern United States, avian nesting success is clearly affected by forest fragmentation. Here, many studies of predation and nest parasitism have shown that the number of species of breeding birds declines much faster from large to small forest patches than from large to small areas within continuous forest. The decline is greatest among forest interior species, insectivorous and predatory species, and long-distance migrants. Nests of birds in small woodlots and near woodlot edges experience much heavier predation than in large forest interior locations. In North America, open-habitat predators such as skunks, crows, jays, and grackles penetrate into the edges of small woodlots. In many cases, the entire forest patch may become accessible to such species. Nest parasitism is also heavy in fragmented forests. In Wisconsin, Indiana, Illinois, and Missouri, for example, eight species of Neotropical migrants show increases in nest parasitism by Brown-headed Cowbirds as forest cover decreases. In regions with less than 55 percent forest cover, one study showed that most Wood Thrush nests contained cowbird eggs. Nest predation also increased as forest cover decreased, with ground-nesting birds experiencing the greatest predation.

The relation of forest fragmentation to predation and cowbird parasitism varies somewhat from region to region. In the northern Rocky

Mountain region of the western United States, for example, nest predation is greater in less-fragmented forests than in more-fragmented areas. This appears in part to be due to the presence of a forest interior predator, the red squirrel, which tends to be absent in smaller forest stands. Generalist predators, such as those implicated in nest predation in many parts of eastern North America, do not seem to concentrate their activity on small stands in this region. Although cowbird parasitism increases in more fragmented forest stands, as it does in eastern North America, the intensity of parasitism is more closely related to the presence of human activity than to forest patch size itself. Livestock pens, feedlots, and houses with bird feeders apparently provide basic food sources for cowbirds. These observations emphasize the importance of the overall landscape context in determining the incidence of nest predation and parasitism in forest patches.

In Europe and Asia fragmentation of forests is substantially greater than in North America (Table 4.2). Although boreal conifer forests in Europe and Asia were somewhat less fragmented than those of North America, temperate forests are substantially more fragmented, especially in Europe. Mediterranean forests in Europe are severely fragmented, mostly due to human activities.

### HABITATS USED IN MIGRATION

Long-distance migrants also face problems of habitat availability and quality along their migration routes. The energy they consume is supplied by fat that is deposited before and during migration. Predeparture staging areas and migration stopover areas where they can refuel are essential during migration both to and from breeding areas. Removal and fragmentation of natural vegetation along migration routes can be detrimental to migrants even if breeding and nonbreeding areas remain favorable. Numerous studies document the use of individual stopover areas, but understanding their overall importance requires analysis of regional landscape conditions. In turn, this requires understanding in considerable detail the patterns of orientation and capability for flights between stopover locations for diverse migratory species.

Modeling studies have evaluated the adequacy of stopover sites for Neotropical migrants in eastern North America during spring migration. These analyses explored various assumptions about distance and direction of orientation during nightly flights, as well as different arrival points of transgulf migrants along the southern coastline of the United States. The adequacy of stopover areas throughout the eastern United States was also included in the models. The results suggest that birds following northeastward migration paths that parallel the Appalachian Mountain chain should encounter suitable stopover areas with little difficulty. Those following northwestward migration paths appear also to encounter suitable stopover sites. In contrast, birds with migration paths directly northward from their arrival point on the central Gulf of Mexico coast do not encounter as favorable a distribution of stopover sites. The analysis thus suggested that major patterns of deforestation and forest fragmentation may now be detrimental to Neotropical migrants following some migration routes.

## The Tropics

The tropics, wintering areas for long-distance migrants, are also affected by the transformation and fragmentation of natural vegetation. In the tropics, forest area is declining by an estimated 580,000 square kilometers, or about 5.2 percent, per decade, according to an analysis for the period 1990–1997. This may well be an underestimate, since estimates for the 1980s were substantially higher. Tropical forests are being lost to clearing for shifting cultivation, creation of pasture and permanent farms, timber harvest, and expansion of urban areas. Along with deforestation, remaining tropical forests are being fragmented into patches of varying size (Table 4.2), for which climatic edge effects and isolation from other patches can be major influences.

The interaction between changing climate and direct human impacts on tropical forests is strong, as well. In humid tropical lowlands, deforestation appears to be a significant driver of climate change. Tropical deforestation is especially likely to act as a positive feedback to climatic warming in areas such as the Amazon Basin of South America. There, deforestation itself might lead to warming by 1.4°C and a major decrease in rainfall.

Estimates of the extent of tropical deforestation vary, depending on criteria for defining forest cover and the methodology employed, such as national inventories as opposed to satellite imagery. Overall, the estimate by the Food and Agriculture Organization of the United Nations (FAO), which defines *forest* as land having tree cover of 40 percent or more, is that 34.7 percent of original tropical forest has been removed. Satellite-based analysis, generally defining *forest* as land having 60 percent or greater tree cover, gives an estimate that 50.1–52.1 percent of tropical forest has been removed. In all cases the remaining forest area includes not only primary forest but partially logged and secondary forest and forest plantations.

Since the 1990s, the overall annual rate of tropical deforestation has increased from 0.57 percent to 0.62 percent (10.40 million hectares), with annual clearing of primary tropical forests rising from 0.66 percent to 0.81 percent (6.26 million hectares).

Deforestation in the New World tropics is less that in other world regions, corresponding to a total of 17.9 percent by FAO estimates, or 33.6–38.2 percent by satellite-based estimates. Central and South America are estimated to have lost 36 and 37 percent, respectively, of original forests. The Amazon Basin of South America, however, is estimated to hold about half of the world's remaining tropical forest.

In the African tropics, deforestation is very severe. According to FAO estimates, only 36.6 percent of original tropical forest has disappeared, but in all likelihood, these estimates of forest include very large areas of dry, fairly open savanna forest that has suffered less than more moist forest has. Satellite-based estimates of closed forest suggest that 64.8–68.7 percent has disappeared.

Tropical Asia is estimated to have lost most of its original forest. Estimates by the FAO suggest that 61.3 percent of original forest has been lost. Satellite-based analyses give roughly similar estimates of loss: 57.3–68.5 percent. Deforestation is particularly high in some parts of southeast Asia, such as Malaysia, where it has recently increased greatly.

Tropical cloud forest, which amounts to about 2.5 percent of the world's tropical forest, is one of the most threatened types of tropical forest. Almost 60 percent of the world's cloud forests are located in tropical Asia, about 25 percent in the Americas, and about 15 percent in Africa. In South America, very little of the original Andean and sub-Andean forests of Colombia are protected. These forests are also highly fragmented.

Fragmentation of tropical forest communities is a major problem in wintering areas of many migratory birds (Table 4.2). The Asian tropics show the highest percentage of forest fragmentation, due almost entirely to human activity. Mangrove forests in this region are more disturbed than elsewhere in the tropics. The tropical forests of northern Queensland, Australia, also show high fragmentation. Fragmentation is very high in the drier tropical forests of Africa, although tropical moist forests show less fragmentation in Africa than in Asia.

In the North American region, tropical forests in southern Mexico and Central America show high levels of fragmentation (Table 4.2). Fragmentation is even substantial in the vast tropical rain forests of South America, with only a third of forest area qualifying as true interior (Table 4.1).

## Southern Hemisphere Temperate Regions

Habitat fragmentation is also a significant problem in temperate regions of the Southern Hemisphere. South African forests show greater fragmentation than those of Australia and South America. The Mediterranean zone forests of Chile show very high fragmentation, and those of Australia and South Africa show substantial fragmentation. In the wheat belt of Western Australia, nature preserves consisting of woodland and shrubland now exist as islands in an intensively farmed landscape.

# Degradation of Arid and Semiarid Lands

Desertification is the reduction of biological productivity of arid or semiarid regions by natural or anthropogenic causes. Over broad geographical areas, such as northern Africa, it is evident as the expansion of areas of vegetation and biological productivity typical of drier climatic zones. Since 1969, the Sahel of sub-Saharan Africa, roughly the region from 13° to 20°N, has experienced much drier conditions than in earlier years. This area is of major importance for many migratory birds of the western Palearctic, both as a wintering area and as an area of migratory passage.

Uncertainty exists about the cause of the long Sahelian drought. Between 1968 and 1997, annual rainfall in this region dropped sharply to a level 25–40 percent lower than in the preceding three decades. The suddenness and consistency of this change suggest that it was due to a switch between alternate stable climatic states. Modeling studies also suggest that the shift from a wetter climate to a drier climate was caused by a strong interaction between climate and vegetation cover. At some critical point, gradual degradation of the plant cover due to human activity, combined with climatic effects of a change in surface temperature of the Atlantic Ocean, appears to explain the suddenness and persistence of the climatic shift. Increased monsoon rainfall is predicted to occur during this century and may improve conditions in this important wintering area for trans-Saharan migrants.

Many species of birds that migrate between the western Palearctic and sub-Saharan Africa have declined in abundance since 1970 (see Chapter 8). Declines are greatest for species wintering in arid open habitats in Africa. Some of these, such as the European Roller, Pallid Harrier, and Lesser Kestrel have declined to the point that their survival is threatened. These declines are consistent with reduced survival of wintering birds, perhaps coupled with poorer feeding conditions at staging and stopover areas during northward migration. The prolonged drought affecting the Sahel may well be leading to increased mortality of passerine migrants crossing the Sahara in spring.

## Alteration of Ecosystem Processes

Worldwide, livestock grazing is a major activity in areas of otherwise natural vegetation. In the western United States, for example, about 70 percent of the overall landscape is grazed by cattle. These lands range from desert scrub to coniferous forest. Studies in several western states have found that vegetational changes induced by grazing favor some bird species and are detrimental to others, both residents and migrants. Long-distance Neotropical migrants, however, were significantly less abundant in grazed than nongrazed areas.

Alteration of fire regimes is still another widespread influence on migratory birds. Human influences on fire are complex and changing. In some world regions, including much of the boreal forest zone, human activities have increased the frequency and severity of wildfire. In other parts of the temperate zones and tropics, past activities of aboriginal peoples and subsistence farmers probably also increased fire frequency. In the last century, however, efforts have been made in most regions to prevent wildfires. In any case, the result is a changing pattern of fire promotion and suppression.

Some species of migratory birds usually benefit from the effects of fire, whereas others do not. In coniferous forest areas in Montana, for example, migratory birds such as Cassin's Vireo and Swainson's Thrush generally declined after wildfire, whereas Townsend's Solitaires and Lazuli Buntings increased. The intensity of wildfires also determines the effect of fire on populations of birds. In the Montana study, several migratory birds increased in abundance in areas of moderate fire intensity but decreased in areas of severe intensity.

Clearly, long-term trends in fire incidence in natural plant communities can alter the regional abundance of migratory bird species. Thus, to the extent that climatic change interacts with the natural and human-caused frequency of wildfire to establish new regimes of fire frequency, decreases or increases in abundance of fire-sensitive species will be expected.

Timber harvesting is an additional cause of fragmentation and hetero-
geneity in natural forest vegetation. In areas like western North America,
for example, harvesting greatly reduces the area of old-growth forests and
the incidence of standing dead trees. Studies of harvested forests in the
north-central Rocky Mountains show that several migratory birds are negatively affected by clear-cutting practices that fragment these forests and increase the extent of edge conditions.

The introduction of alien animals and plants to new geographical regions is a pattern of global scope. The European Starling, introduced to
North America in 1890, has spread throughout most temperate portions of
the continent. Although it is primarily a cavity nester, it seems to have had
little long-term effect on many native cavity-nesting birds. Three species of
cavity-nesting sapsuckers have decreased following arrival of starlings, however, possibly reflecting competition for nest cavities.

The Brown-headed Cowbird, a native North American species that has
greatly expanded its range because of human activity, is a nest parasite on
more than 200 species of songbirds. Once a nomadic species that followed
herds of bison, it is now more widespread and sedentary, putting its hosts at
risk over more of their breeding season. Cowbirds can seriously reduce the
reproductive success of many hosts with which they have come into contact
relatively recently. In the 1960s, for example, about 55 percent of the nests
of Kirtland's Warbler were parasitized by cowbirds. At that time this warbler was critically endangered. Recent cowbird control efforts have reduced
this parasitism considerably.

More recently, cowbird parasitism has become a serious problem in the
American Midwest, where forest is severely fragmented. Cowbirds forage
in farmland but parasitize forest birds such as the Wood Thrush and Veery.
Some regional populations of these migratory thrushes are apparently population sinks that are maintained by dispersal from other regions. Cowbirds
also parasitize many migratory warblers, vireos, tanagers, flycatchers, and
orioles.

Alien plants, introduced deliberately or inadvertently, have transformed natural plant communities throughout the world. In many cases,
this has been detrimental to native migratory birds. On the Columbia
Plateau in the western United States, for example, annual cheatgrass has replaced native sagebrush over vast areas. Few native birds, especially migratory species, use cheatgrass habitat. Those that decline when sagebrush is
converted to cheatgrass include the Sage Thrasher, Sage Sparrow, Brewer's
Sparrow, and a number of other species. On the other hand, Burrowing

Owls and Long-billed Curlews seem to increase when cheatgrass replaces sagebrush, apparently benefiting from the more open habitat.

In western North America, salt cedars, Russian olives, and other alien shrubs and trees have invaded and often replaced native riparian woodland or shrubland. These invasions are detrimental to many migratory bird species, although some species have benefited.

Alien animals represent another serious threat to migratory birds, especially land and sea birds on oceanic islands. Herbivorous mammals can decimate native vegetation, and predators such as cats and rats can extirpate populations of migratory seabirds.

## Chemical Pollution

Acid precipitation remains a serious form of pollution in areas downwind from large urban and industrial areas throughout the world. In some regions, such as eastern Asia, acid precipitation is becoming worse.

In North America and Europe, damage to high-elevation forests by acid precipitation is extensive. Declining populations of migratory birds are associated with several of these areas. The most serious population declines of passerine birds in North America over the period 1966–1992, for example, were concentrated in areas of the Adirondack Mountains, the Blue Ridge Mountains, and the Cumberland Plateau that were subject to acid precipitation.

For the Wood Thrush, a Neotropical migrant, population trends appear to be related to acid precipitation. Data on thrush abundance from Cornell University's Birds in Forested Landscapes study and the North American Breeding Bird Survey, together with data on acid deposition and soil acidity, show that a strong negative correlation exists between acid precipitation and Wood Thrush breeding success. This relationship is strongest at high elevations and in the most fragmented forest areas. The most likely cause appears to be reduction in calcium availability due to acidified soil conditions. Wood Thrushes very likely utilize calcium-rich prey organisms, which decrease in abundance in these acidified forests. In the Netherlands, as well, eggshell thinning in woodland insectivorous birds appears to be related to reduced calcium availability due to acidification of the forest ecosystem.

Food chain effects of the acidification of aquatic ecosystems have led to declines of populations of several waterbirds. Fish-eating migratory birds

such as Common Loons and Common Mergansers show reduced reproductive success in lakes that are acidic and have lost fish. Many migratory
waterfowl, such as the American Black Duck, also feed on aquatic invertebrates during the breeding season, and declines of some of these species
may be related to decreased invertebrate populations in acidified waters.
Some migratory songbirds, including swallows and flycatchers that nest
near water and depend on aquatic insects, are also affected. These species
are reported to show eggshell thinning and reduced reproductive success in
areas with acidified waters.

Other pollutants, such as mercury, introduced via combustion of fossil
fuels, may also have negative effects on migratory birds. High levels of mercury, for example, occur in Bicknell's Thrush, a Neotropical migrant nesting at high elevations in New England and adjacent Canada and wintering
in the West Indies. Waste discharges of chemical nutrients, particularly nitrogen and phosphorus, lead to changes in the productivity and biotic composition of aquatic environments. This effect, eutrophication, alters food
resources of many aquatic birds.

## Ultraviolet Radiation

Depletion of the stratospheric ozone layer over polar regions has led to
concern that high levels of ultraviolet radiation could cause catastrophic destruction of plankton-based marine food chains, especially in Antarctic waters. Populations of penguins and other seabirds, which depend on these
food chains, might therefore be affected. Planktonic organisms show accli-
mation to ultraviolet, however, and damage from current levels of radiation
may be much less than feared. Nevertheless, the possibility exists that migratory seabirds might be affected by disruption of food chains in ocean
areas receiving intense ultraviolet radiation.

## Summary

Climatic change interacts with many other influences of human activity on
the global environment. Foremost are the destruction and fragmentation of
natural plant communities. Increasingly, however, intensified land and water management practices, introduction of alien species, and chemical pollution are degrading the habitats used by migratory birds throughout their
annual cycles.

## KEY REFERENCES

Achard, F., H. D. Eva, H.-J. Stibig, P. Mayaux, J. Gallego, T. Richards, and J.-P. Ma-lingreau. 2002. "Determination of deforestation rates of the world's humid forests." *Science* 297:999–1002.

FAO. 2005. Global forest resources assessment 2005. FAO Forestry Paper 147. Food and Agriculture Organization of the United Nations, Rome. http://www .fao.org/forestry/site/fra/en.

Foley, J. A., M. T. Coe, M. Scheffer, and G. L. Wang. 2003. "Regime shifts in the Sahara and Sahel: Interactions between ecological and climatic systems in northern Africa." *Ecosystems* 6:524–539.

Hames, R. S., K. V. Rosenberg, J. D. Lowe, S. E. Barker, and A. A. Dhondt. 2002. "Adverse effects of acid rain on the distribution of the Wood Thrush *Hylocichla mustelina* in North America." *Proceedings of the National Academy of Sciences USA* 99:11235–11240.

Hejl, S. J., D. E. Mack, J. S. Young, J. C. Bednarz, and R. H. Hutto. 2002. "Birds and changing landscape patterns in conifer forests of the north-central Rocky Mountains." *Studies in Avian Biology* No. 25:113–129.

Koenig, W. D. 2003. "European Starlings and their effect on native cavity-nesting birds." *Conservation Biology* 17:1134–1140.

Rimmer, C. C., K. P. McFarland, D. C. Evers, E. R. Miller, Y. Aubry, D. Busby, and R. J. Taylor. 2005. "Mercury concentrations in Bicknell's Thrush and other in-sectivorous passerines in montane forests of northeastern North America." *Eco-toxicology* 14:223–240.

Robinson, S. K., F. R. Thompson III, T. M. Donovan, D. R. Whitehead, and J. Faaborg. 1995. "Regional forest fragmentation and the nesting success of migratory birds." *Science* 267:1987–1990.

Smucker, K. M., R. L. Hutto, and B. M. Steele. 2005. "Changes in bird abundance after wildfire: Importance of fire severity and time since fire." *Ecological Applica-tions* 15:1523–1549.

Wade, T. G., K. H. Riitters, J. D. Wickham, and K. B. Jones. 2003. "Distribution and causes of global forest fragmentation." *Conservation Ecology* 7(2):7. http:// www.consecol.org/vol7/iss2/art7.

Whitcomb, R. F., C. S. Robbins, J. F. Lynch, B. L. Whitcomb, M. K. Klimkiewicz, and D. Bystrak. 1981. Effects of forest fragmentation on avifauna of the eastern deciduous forest. Pp. 125–205 in R. L. Burgess and D. M. Sharpe (Eds.), *For-est Island Dynamics in Man-dominated Landscapes*. Springer-Verlag, New York.

Wright, S. J. and H. C. Muller-Landau. 2006. "The future of tropical forest species." *Biotropica* 38(3):287–301.

# PART III

---

# Ecological Responses of Migratory Birds to Global Change

In this section we shall begin by considering major avian habitats that are especially threatened by changing climate. We shall then consider how major groups of birds are being affected by climatic change. First, we shall survey major regional migration systems of land birds. Then, we shall examine migration patterns of specific groups of land, freshwater, and marine birds.

# Chapter 5

# *Physical and Biotic Challenges to Migratory Bird Responses*

Human activities are thus changing climatic patterns, habitat conditions, and resources that clearly affect the distribution and seasonal activity patterns of many species of plants and animals. Migratory birds are certain to be affected in many ways. In some cases, entire habitat types essential for breeding, migration stopovers, or nonbreeding use are being degraded, reduced in extent, or completely transformed. In many other cases, climatic change is lengthening, shortening, or altering the routes along which migratory birds must move.

In this chapter we shall examine habitat and food resource patterns that make some birds more vulnerable to global change than others. We shall also identify general factors that constrain the response of migratory birds, as species and as members of bird communities, to these changing conditions.

## Bird Habitats at High Risk from Global Warming

Global climatic change might eliminate certain habitats, leading to extinction of local or global populations of migratory birds. In mountain regions, for example, climatic warming might eliminate alpine habitats accessed for breeding by migratory species. In the high Arctic, tundra habitats bordering the Arctic Ocean are likely to shrink or disappear as climatic warming allows forests and shrublands to move northward. Thawing of the permafrost is already modifying the hydrology of tundra ponds and lakes. Warming is also affecting nesting habitats and marine foods of migratory birds in the Antarctic. Rising sea levels due to global warming might eliminate coastal and insular breeding areas, migratory stopover sites, or wintering habitats. Interior wetlands in continental areas are also likely to diminish in area because of warmer and drier climates.

## Bird Habitats of High Mountains

High-mountain ecosystems are at particular risk of disappearing in the face of global warming. Climatic warming is altering not only the physical conditions of high mountains but also basic aspects of plant ecology. Ecosystems such as alpine tundra and tropical cloud forest may gradually be displaced upward until they disappear. In alpine areas of the Austrian Alps, for example, the vegetation above 2900 meters is being invaded by plants from lower elevations. In Sweden, the tree line has risen by 100 to 150 meters since the 1950s, and a number of shrub and tree species have colonized alpine snow-bed communities. Although these are not noxious weeds, they do signal the fact that the ecology of alpine areas is undergoing change.

Species that breed in alpine environments experience great year-to-year variation in weather and reproductive success. In summers with severe weather conditions, for example, only a small percentage of female alpine Willow and White-tailed Ptarmigan are successful in producing broods. Along with low reproduction rates, many of these birds also have long generation times. Thus, the evolutionary responses of birds of alpine areas to gradual climate change may be slow, and rapid climatic change is likely to translate into population declines.

Most birds of high-mountain environments are altitudinal or latitudinal migrants. The rosy finches of western North America and mountain finches of Asia are examples of high-mountain birds that are at risk from loss of alpine tundra habitat. Black Rosy Finches and Brown-capped Rosy Finches nest at elevations above 3050 meters in the Great Basin and southern Rocky Mountains of the western United States. Near Santa Fe, New Mexico, where I live, the Brown-capped Rosy Finch once nested at an elevation of about 3506 meters in the scree slope above Nambe Lake, a cirque lake below Lake Peak in the Sangre de Cristo Mountains. To see this species and the other North American rosy finches, however, we annually make a winter pilgrimage to Sandia Crest, at 3255 meters in the mountains above Albuquerque. Here we can watch three species of rosy finches visit feeders at a small café and gift shop, and even watch the birds being banded. Climatic warming may already have caused the Brown-capped Rosy Finch to abandon its breeding site in the Sangre de Cristos.

As we have seen, tropical cloud forest, already under severe threat from deforestation, is one of the world's most threatened ecosystems. Some, if not many, cloud forest birds are altitudinal migrants, as well, and thus depend not only on cloud forests for breeding but on lower-elevation communities for other phases of their life cycle (see Chapter 10). In Costa Rica, warming and drying of the cloud forest zone in the Tilarán Mountains between 1979 and 1998 has enabled species of foothill birds to become established at cloud forest elevations. Many other foothill species have appeared as visitors or short-term residents. These species are diverse in their taxonomy and ecology, and some are likely to have negative effects on montane species.

Mountain forest areas in Africa, northern Australia, and southeast Asia are also warming, enabling birds of lowland areas to move to higher elevations. One study revealed that between 1971 and 1999, ninety-one species of common breeding birds in southeast Asia shifted their upper altitudinal range limits to higher elevations, compared with thirty-four species that showed a drop in their upper limits. This region has warmed by at least 0.3°C over this period and is projected to warm by 1.1°C to 4.5°C by AD 2070. Although altitudinal migration patterns of high-montane birds in these regions are poorly known, it is likely that the invasion of birds from lower elevations will cause some reductions in their populations.

As climate warms, high-mountain ecosystems are also likely to become both reduced in area and increasingly isolated from each other. Thus, their bird populations will become increasingly subject to stochastic fluctuations typical of fragmented populations. In the tropical forests of northeastern Australian mountains, for example, modeling studies showed that the habitat area for endemic vertebrates would be reduced catastrophically by predicted climatic warming. For a climatic warming of 1.0°C, about 40 percent of the core habitat of endemic species would be lost, and if warming reached 3.5°C, the loss would be about 90 percent.

Warming also allows predators, parasites, diseases, and competitor species to spread to higher elevations. In Hawaii, for example, with climatic warming, mosquitoes carrying avian malaria are likely to spread upward into forests occupied by native birds, most of which lack resistance to the disease. This may have already contributed to the endangerment and extinction of some native birds. Further warming by 2.0°C will mean that all highland forests will become vulnerable to avian malaria.

## Bird Habitats of High Latitudes

Species that breed at high latitudes are also likely to experience major changes in their breeding habitat with climatic warming. Expansion of vegetation types adapted to warmer climates is reducing areas of Arctic and high-latitude alpine tundra. On the North Slope of Alaska, for example, the area covered by large shrubs is increasing at a rate of 1.2 percent per decade. The tree line is also advancing into tundra areas. Over the past 50 years, spruce forest has replaced an estimated 11,600 square kilometers of tundra. At the same time, coastal erosion due to rising sea level is tending to reduce tundra area still further.

Modeling studies show the effects of a continued global warming of 2°C on Arctic ecosystems, which will likely occur between AD 2026 and 2060. Arctic tundra is projected to decrease by about 42 percent under the most likely scenario, being replaced by boreal forest. Two-thirds of Alaskan tundra would likely be converted to forest. Forest communities will potentially advance northward by up to 400 kilometers, except as limited by slow tree dispersal rates.

Some migratory bird species with breeding ranges at very high latitudes may be at risk of extinction if the warming climate eliminates their habitat or alters habitat conditions in ways that enable competitor species to invade and displace them. In particular, northward retreat of Arctic tundra environments may pose an especially serious threat to many tundra-breeding shorebirds.

The seasonal dynamics of Arctic tundra lakes, ponds, and streams are also being altered. The ice-free season for ponds and lakes is being extended, and evaporation losses are increasing. As we noted in Chapter 3, many of these ponds and lakes are disappearing because of the thawing of the permafrost.

At high latitudes in the Antarctic region, some species are also at risk from climatic warming. The Antarctic Peninsula has experienced perhaps the greatest warming of any location on earth. This warming is correlated with southward retreat of populations of some penguins and advances by others. We shall consider this case in detail in Chapter 16.

## Ocean Coastlines and Atolls

In coastal areas and low-lying oceanic islands, rising sea level may reduce or eliminate critical breeding or nonbreeding habitat. In the Pacific and Indian oceans, numerous atolls, with elevations only a few feet above mean sea level, are at risk of disappearing.

The northwestern Hawaiian Islands, low-lying atolls used as nesting areas by seabirds and other marine animals, are already affected by rising sea level. Whale Skate Island, one of the small islands that form French Frigate Shoals, is an example of how seabird habitat can be lost. This island, about 4 to 6 hectares in size, was vegetated and supported colonies of nesting seabirds, as well as Hawaiian monk seals and green sea turtles. Erosion, apparently due to rising sea level, has caused the island to disappear over the course of 20 years. More serious erosion may occur if ocean acidification, due to an increase in atmospheric $CO_2$ concentration from the current level of over 387 parts per million (ppm), rises to about 480 ppm. At that point, carbonate deposition in coral reefs will essentially cease, making them even more susceptible to damage by rising seas. We thus stand at a point about halfway between the preindustrial level of 280 ppm and a level critical to the reef-building organisms that maintain coral atolls that are the nesting areas for many seabirds.

Concern also exists over the effects of rising sea levels on estuary habitats used by breeding and migrating shorebirds. On the west coast of the United States, even a conservative estimate of climate warming will result in a 20 to 70 percent loss of intertidal habitat used for foraging by shorebirds. Substantial habitat alterations are also likely along the Atlantic and Gulf coasts of the United States. The salt marshes of Jamaica Bay, New York, have lost about 12 percent of their area and 38 percent of their vegetation since 1959. In England, on the other hand, current efforts to model habitat changes due to sea level rise suggest that loss of muddy and sandy habitats will not be extensive enough to reduce numbers of migrating or wintering shorebirds. Clearly, too, the effects of rising sea level will vary greatly with coastal geomorphology.

## Wetlands in Arid Regions

Increasing aridity in areas such as central North America may reduce the extent of aquatic habitats utilized by shorebirds and waterfowl. For the Prairie Pothole Region of the northern United States and south-central

Canada, for example, a wetland simulation model was used to analyze the potential changes in waterfowl habitat. This region is the most productive waterfowl area in the world, and it supports from 50 to 80 percent of North America's duck populations. Even in this productive region, duck populations are notoriously variable from year to year because of variations in rainfall that affect the number of breeding ponds.

The wetland model predicts that drought conditions will intensify in the Prairie Pothole Region during the coming century. Based on data from eighteen locations and 95 years of weather records, simulations suggest that the best breeding habitat for waterfowl will shift from the Dakotas and southeastern Saskatchewan to more eastern and northern portions of the Prairie Pothole Region. The latter areas, although wetter at present, are now less productive of waterfowl, in part because of drainage of many wetland areas. Whether or not population shifts can compensate to maintain current waterfowl production levels is thus uncertain.

Wetlands in arid regions are also important breeding and migratory stopover areas for shorebirds. Shorebirds in general are suffering population declines; thirty of the fifty-six North American species are in decline. In particular, those species that migrate through central North America have shown greater population declines than those following coastal routes. It is likely that the declines of these interior migrants are at least in part due to loss of wetland habitats used in migration. In Europe, the decline in numbers and shift in breeding range of the Ruff, a wetland-breeding shorebird, is at least partly due to climatic change.

## Food Resources at High Risk from Global Warming

Seabirds are at particular risk from changes in ocean current systems and productivity patterns due to climate change. For these species, access to ocean areas rich in food resources is essential throughout the annual cycle. Altered patterns of nutrient availability in marine ecosystems are affecting food resources critical to many marine birds. Changes in the periodicity of the El Niño/La Niña cycle in the tropical Pacific Ocean, for example, influence nutrient upwelling and fish production in the Galápagos Islands, along the west coasts of North and South America, and even in subantarctic waters.

The El Niño of 1997–1998 illustrates how severely this event can affect seabird populations. In 1996, populations of Guanay Cormorants, Peruvian Boobies, and Peruvian Pelicans had peaked at 6.8 million birds. The El

Niño of 1997–1998 decimated populations of anchoveta and other fish on which seabirds depended, and it caused bird numbers along the Peruvian coast to crash. Many of the birds moved south to southern Peru and Chile, but nest desertion was nearly complete, and mortality of immature birds and adults was heavy. In late 1997 the numbers of Guanay Cormorants and Peruvian Pelicans on the Peruvian coast were less than 1 percent of their 1996 numbers, and Peruvian Boobies numbered only about 13 percent of those in 1996.

In the northeastern Atlantic Ocean, warming waters have led to changes in phytoplankton communities and the distribution of marine fish. In some areas with cooler waters, plankton organisms have shifted northward, and their overall abundance has increased with warming of the ocean. In areas with waters that are already fairly warm, additional warming has decreased their abundance. In the North Sea, two-thirds of fish species have moved northward, their northern limits shifting by 119 to 816 kilometers. We shall examine the influences of these changes on sea birds in detail in Chapter 14.

The changing sea ice patterns of both the Bering Sea and Antarctic Ocean also influence food resource abundance for migratory birds. In the Bering Sea, northward retreat of the ice pack is leading to change in food resources for birds and mammals. In ice-dominated areas, mollusks and other benthic invertebrates predominate, and vertebrates such as Spectacled Eiders and marine mammals are the major predators. With ice retreat, invertebrate-based food chains are replaced by those based on pelagic fish.

Along the northern Antarctic Peninsula, warmer weather and increased melting of glacial ice has created a lens of low-salinity water that extends to depths as great as 25 meters and for a distance of up to 100 kilometers offshore. This water regime favors the dominance of producer organisms less than 10 micrometers in size, rather than larger diatoms and other colonial algae. These producers are most efficiently grazed by salps, gelatinous chordates that resemble jellyfish, rather than shrimplike krill. Salps, in turn, are nutritionally poor and not preferred foods of planktivorous penguins and other seabirds. With the decline in area of sea ice around Antarctica, the phytoplankton that support krill thus have declined and salps have increased in relative abundance over large areas of the Southern Ocean. Blooms of the larger phytoplankton now occur farther south, as do the concentrations of krill and major krill predators such as the Adélie Penguin. These changes also have the potential for seriously reducing populations of albatrosses, seals, and whales that also feed primarily on krill.

## Evolutionary Constraints to Response by Migratory Birds

It is important to consider whether or not migratory birds can show evolutionary adjustments quickly and fully to the climate changes that are now occurring. Many temperate zone birds do seem to be showing changes in migration patterns in response to climatic change. As we shall see, these include abilities to tolerate changes in habitat conditions, to adjust the frequency of migrant and resident individuals within populations, and to alter migration schedules and routes. But some major constraints exist.

### Constraints Based on Habitat Specialization

As we noted earlier, many biologists have suggested that migratory birds are at special risk from climate change because they depend on conditions in different geographical areas and often distinctly different habitats at different times of year. There are few general tests of this idea, however. One study analyzed factors correlating with population changes among breeding land birds in France. Of the species considered, twenty-two were trans-Saharan migrants, thirty-two partial or short-distance migrants, and twenty-three residents. This analysis suggested that species with specialized habitat requirements and those with more-northerly distributions in the country, whether migratory or not, were most prone to declines. Among migratory Old World warblers of the genus *Phylloscopus*, for example, species with greater habitat specialization showed the greater declines (Figure 5.1).

### Constraints to Changes in Migration Tendency

Many areas of the North Temperate Zone and Arctic are experiencing milder winters, while summers are remaining the same or warming to some degree. Many populations of birds of these regions are partial migrants. In North America, I estimate, about 43.6 percent of the migratory species that breed in the Nearctic region are partial migrants. In South America about 158 species are partial migrants, which corresponds to 59.4 percent of all austral migrant species. In Europe, an estimated 57 percent of all breeding bird species are partial migrants. With climatic change, it appears that partially migrant populations might be able to shift quickly to permanent residency or fully migrant status. Many short-distance migrants might also

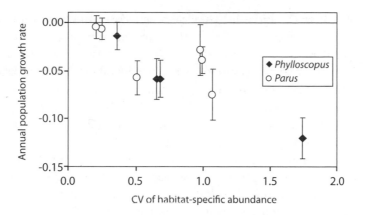

FIGURE 5.1. Habitat specialization, measured as the coefficient of variation (CV) of habitat-specific abundance, among six nonmigratory tits (*Parus* spp.) and four migratory Old World *Phylloscopus* warblers in relation to recent population change. From left to right the warblers are Common Chiffchaff, Western Bonelli's Warbler, Willow Warbler, and Wood Warbler. (From Julliard, R., F. Jiguet, and D. Couvet. 2004. "Common birds facing global changes: What makes a species at risk?" *Global Change Biology* 10:148–154.)

have the genetic variability to permit evolutionary change to partial migrants, full migrants, or permanent residents. Long-distance migrants, on the other hand, may lack the evolutionary ability for such changes.

Partial migration is also shown by many tropical birds, especially in mountainous regions. Altitudinal movements can occur in response to short-term periods of unfavorable weather at high elevations or longer periods of unfavorable resource availability due to seasonal change. Many of the breeding birds of cloud forest areas from southern Mexico through South America, for example, carry out altitudinal movements. In addition, many fruit-eating and insectivorous birds of tropical lowlands show extensive movements correlated with seasonal changes in food availability.

Changes in seasonal conditions, particularly the appearance of favorable weather earlier in the spring, should permit many short-distance and altitudinal migrants to alter their migration schedules quickly. Adaptive responses of this sort are frequent in so-called "weather migrants" that move in close response to favorable or unfavorable conditions. On the other hand, in mountainous regions, the effects of climatic change can differ markedly at different elevations. At the Rocky Mountain Biological Laboratory, at an elevation of 2945 meters in Colorado, USA, an increase in snowfall at high elevations has paralleled the increase in warmth at low elevations, so the appearance of thawed ground has not changed over time. American Robins, apparently in response to the earlier development of warm weather at lower elevations, were arriving about 14 days earlier in spring in 1999 than in 1981. Thus, the time between robin arrival and the first appearance of bare ground has increased by about 18 days.

For long-distance migrants, in contrast, the timing of migration appears to be strongly genetic, and their response is likely to require evolutionary adaptation. In addition, to the extent that global warming leads to a conversion of partial migrants into permanent residents and short-distance migrants into partial migrants, long-distance migrants may become disadvantaged. If the resources of their breeding habitats become more fully utilized by residents or early-arriving short-distance migrants, long-distance migrants may suffer from intensified competition.

## Constraints to Change in Migration Routes

The migration routes used by some long-distance migrants appear to have an evolutionary basis, and these migrants tend to be conservative in their response to changes in breeding or nonbreeding ranges. Pectoral Sandpipers, which have a wide breeding range across Arctic North America, also breed in the Siberian Arctic west to the Taymyr Peninsula (longitude 110° E). The migration pattern of most of these sandpipers from their wintering range in South America follows a route north through the Americas to the Arctic, and then west to Siberia. Some of these birds do winter along the eastern Asian coast. Several Eurasian land birds, including the Bluethroat, Arctic Warbler, Northern Wheatear, Western Yellow Wagtail, and White Wagtail, have extended their breeding ranges into western Alaska, but in fall they migrate west across the Bering Strait to eastern Asia and from there to wintering ranges in southern Asia and even Africa.

A similar situation exists for some birds that winter in the Neotropics. The bulk of Swainson's Thrushes that winter in northern South America and breed in Alaska and Canada follow a northward migration route that crosses the Caribbean Sea and Gulf of Mexico into the eastern United States. These migrants then turn west to reach the far western portion of their breeding range in the coniferous forest region of North America. For these Old and New World species, the migration route appears to retrace the historical route of invasion rather than following a shorter route to a wintering area seemingly similar in climate to that presently occupied. A significant evolutionary constraint to change in migration route and non-breeding range thus appears to exist for some species.

Climatic change is now altering the distance between breeding and nonbreeding ranges for many species. In the Old World, for example, expansion of areas such as the Sahara, due to climatically induced desertification, increases the distance that Palearctic–African migrants must traverse with very limited access to food and water. Whether or not long-distance migrants can quickly adjust to such changes in ranges and migration routes is uncertain.

## Community Constraints to Response by Migratory Birds

Climate change also has the potential to change the competitive balance among species with different migration strategies, leading to community reorganization. In the Lake Constance region of central Europe, where winter temperatures warmed significantly between 1980–1981 and 1990–1992, the abundance of long-distance migrants relative to short-distance migrants and permanent residents has decreased. A broader examination of European bird communities at twenty-one different locations compared their composition in migrant and resident species in 1972–1976 with that in 1988–1992. This analysis also indicated that with climatic warming, a slight decrease occurred in the proportion of long-distance migrants and a significant increase was shown in the proportion of short-distance migrants. These changes agreed with predictions of models relating climate change to biodiversity.

As we shall see, many migratory species seem already to be responding to climatic change by shifting boundaries of their breeding, nonbreeding, and migratory ranges or by adapting through phenotypic plasticity or evolutionary change. All of these changes alter the pattern of seasonal change in bird communities noted above. In reality, both individual adjustments and evolutionary responses are likely to occur in different degrees in different bird taxa. In some regions, much of the response of the bird assemblage appears likely to be due to changes within individual species populations in the proportions of migratory and resident individuals. In fact, the change in climate now foreseen in most of Europe seems likely to lead mostly to changes in migratory tendency of existing species rather than invasion of nonmigratory species from other regions.

How plant communities change in structure with climatic change will also influence the response of migratory birds. The limited ability of long-lived woody plants to migrate quickly in response to climatic shifts may also lead to a substantial delay in vegetational response that correspondingly affects the responses of migratory birds. Invasive alien plants may thus be quick to invade habitats that have been altered by climatic change. The vegetational structure that results may or may not be suitable for bird species for which the climate of the region has become suitable. Models that project future change in vegetation in response to climate change do not adequately consider such lags.

On the other hand, many of the members of biotic communities do seem to be responding to changing climate in both space and time. A meta-analysis of the responses of many different groups of plants and animals to climate change found that significant poleward shifts in range boundaries of diverse organisms had occurred, averaging 6.1 kilometers per decade, and that significant advances of spring phenological events had also occurred, averaging 2.3 days per decade. Data for 168 species of birds were included in this analysis. Of these, 78 species showed responses consistent with the above trends, 14 opposite to these trends, and 76 without a clear direction. Although not all of these bird species were migratory, it is clear that global change is substantially altering avian biogeography and the structure of the communities birds occupy. Nevertheless, with climatic change, it appears unlikely that entire biotic communities will shift in space and time without significant change in structure. We shall examine the influences of changing climate on bird species and communities in detail in the chapters that follow.

With this introduction to general patterns of influence of environmental conditions on migratory birds, we shall now turn to an examination of how migration patterns are changing in different world regions, and in different groups of birds.

## Summary

Several important bird habitats are at special risk from global warming, including high-mountain areas such as alpine tundra and tropical cloud forest, Arctic tundra, wetlands of dry continental interior areas, and coastal habitats of continental and insular areas. Warming ocean waters are also altering food chains on which oceanic birds and other vertebrates depend. These changes challenge the adaptive capabilities of birds, particularly those with specialized habitat requirements. Nevertheless, several general studies indicate that many migratory species are shifting their range limits in an adaptive fashion and that bird communities are experiencing changes in the balance of resident, migratory, and partially migratory species.

## KEY REFERENCES

Austin, G. E. and M. M. Rehfisch. 2003. "The likely impact of sea level rise on waders (Charadrii) wintering on estuaries." *Journal of Nature Conservation* 11:1–16.

Baker, J. D., C. L Littman, and D. W. Johnston. 2006. "Potential effects of sea level rise on the terrestrial habitats of endangered and endemic megafauna in the northwestern Hawaiian Islands." *Endangered Species Research* 2:21–30.

Julliard, R., F. Jiguet, and D. Couvet. 2004. "Common birds facing global changes: What makes a species at risk?" *Global Change Biology* 10:148–154.

Lemoine, N., W. Jetz, and K. Böhning-Gaese. 2007. "Impact of climate change on migratory birds: Community assembly versus adaptation." *Global Ecology and Biogeography* 16:1–12.

Lindström, Å. and J. Agrell. 1999. "Global change and possible effects on the migration and reproduction of arctic-breeding waders." *Ecological Bulletins* 47:145–159.

Moline, M. A., H. Claustre, T. K. Frazer, O. Schofield, and M. Vernet. 2004. "Alteration of the food web along the Antarctic Peninsula in response to a regional warming trend." *Global Change Biology* 10:1973–1980.

Montes-Hugo, M., S. C. Doney, H. W. Ducklow, W. Fraser, D. Martinson, S. E. Stammerjohn, and O. Schofield. 2009. "Recent changes in phytoplankton communities associated with rapid regional climate change along the western Antarctic Peninsula." *Science* 323:1470–1473.

Parmesan, C. and G. Yohe. 2003. "A globally coherent fingerprint of climate change across natural systems." *Nature* 421:37–42.

Pounds, J. A., M. P. L. Fogden, and J. H. Campbell. 1999. "Biological response to climate change on a tropical mountain." *Nature* 398:611–615.

Ruegg, K. and T. B. Smith. 2002. "Not as the crow flies: An historical explanation for circuitous migration in the Swainson's Thrush (*Catharus ustulatus*)." *Proceedings of the Royal Society of London B* 269:1375–1381.

Williams, S. E., E. E. Bolitho, and S. Fox. 2003. "Climate change in Australian tropical forests: An impending environmental catastrophe." *Proceedings of the Royal Society of London B* 270:1887–1892.

# Chapter 6

# *Northern Hemisphere Land Birds: Short-distance Migrants*

Many of the migratory birds of temperate and Arctic regions of North America and Eurasia are partial or short-distance migrants. Since the breeding and nonbreeding ranges of these birds are close together, climate changes on regional to continental scales should lead to quick response of such species if their migration cues are closely tuned to weather conditions. In this chapter we shall only consider nonraptorial land birds.

In both North America and Europe, short-distance migrants appear to have responded to climatic change. In many locations the timing of their movements has changed, the relative abundance of migrant and resident individuals in local populations has changed, breeding and wintering ranges have shifted, and population densities have altered. In other areas, little or no change in migration patterns has occurred.

## Spring Migration: North America

Analyses of changes in timing of spring migration by North American short-distance migrants have given mixed results. Several clearly indicate that many short-distance migrants are sensitive to weather conditions in spring, tending to arrive somewhat earlier in years with unusually warm spring weather. Several also show that some species have advanced their spring arrivals over the past century, in several cases by 1–2 weeks. Considerable variation exists, however, in part due to geographical location and in part due to techniques of analysis. Analyses based on first-arrival dates can be misleading because declines in abundance of a species tend to reduce the chance of detecting early arrivals. Measures that reflect mean migration periods appear to be less biased.

In the northeastern United States, spring warming is now occurring earlier. In Concord, Massachusetts, for example, plant phenology has advanced about a week since the mid-1800s, reflecting spring temperatures that are now about 2.4°C warmer. At the Manomet Center for Conservation Sciences on Cape Cod Bay, Massachusetts, spring temperatures warmed by 1.5°C just between 1970 and 2002. This warming is correlated with earlier spring arrivals of some short-distance migrants, as measured by the mean date of capture of birds for banding. Over this period, Eastern Towhees have advanced their mean migration dates by 6.8 days, Swamp Sparrows by 9.7 days, and Blue Jays by 11.8 days. On the other hand, several other short-distance migrants have shown no significant advances in mean migration dates. Two species, the American Robin and Hermit Thrush, also showed significant tendencies for earlier spring arrival during positive phases of the North Atlantic Oscillation (NAO), when mild, wet conditions prevail in New England.

In Chicago, Illinois, data for 1979–2002 on short-distance migrants killed in spring migration when they struck windows at McCormick Place were analyzed to see whether the first arrivals, the start of the main migration period, or the median migration date had changed. No significant trends in spring temperatures in the Chicago area were evident during this period. However, all eleven species for which adequate numbers were available responded by earlier migration in years when warmer spring weather prevailed. This pattern was somewhat stronger for males than for females.

Other studies have examined first-arrival dates, and these show varying patterns. A Massachusetts study reported that first arrivals of Chipping Sparrows were earlier in 2002 than in 1970, but several other short-distance migrants showed no advance in arrival. In Maine, only two species of short-distance migrants, the White-throated Sparrow and Red-winged Blackbird, arrived earlier in 1994–1997 than in 1899–1911. In contrast, eight species seemed to be arriving later, and many others showed no significant change in arrival date. Because nearly a century elapsed between the Maine observations and substantial landscape changes occurred, it is difficult to evaluate these data.

In the Cayuga Lake Basin in New York and in Worcester County, Massachusetts, the first spring arrival dates for birds were recorded by bird club

TABLE 6.I. First-arrival dates of short-distance migrants in Cayuga Lake Basin, New York, for which spring arrivals had advanced substantially.

| Species | First Arrival Dates | |
| --- | --- | --- |
| | 1903–1950 | 1951–1993 |
| Yellow-bellied Sapsucker | 30 March | 15 February |
| Eastern Phoebe | 26 March | 15 March |
| Tree Swallow | 5 April | 19 March |
| Ruby-crowned Kinglet | 10 April | 7 March |
| Eastern Bluebird | 10 March | 26 January |
| Hermit Thrush | 8 April | 9 March |
| Gray Catbird | 1 May | 12 March |
| Brown Thrasher | 26 April | 24 March |
| Yellow-rumped Warbler | 17 April | 30 January |
| Common Yellowthroat | 1 May | 13 April |
| Chipping Sparrow | 3 April | 12 March |
| Field Sparrow | 30 March | 11 February |
| Vesper Sparrow | 1 April | 23 March |
| Savannah Sparrow | 31 March | 11 March |
| Fox Sparrow | 25 March | 9 March |
| Lincoln's Sparrow | 11 May | 4 May |
| White-crowned Sparrow | 2 May | 8 March |
| Brown-headed Cowbird | 17 March | 14 January |

Source: Butler, C. 2003. "The disproportionate effect of global warming on the arrival dates of short-distance migratory birds in North America." Ibis 145:484–495.

members for nearly a century. These suggest that many species arrived earlier in 1951–1993 compared with 1903–1950. Of thirty-four species of short-distance migrant songbirds, twenty-seven showed earlier arrival in Massachusetts, twenty-six in New York, and twenty-two in both areas (see Table 6.1 for some New York data). Species not advancing their arrival included some vireos, warblers, and sparrows. On average, short-distance migrants arrived 13 days earlier in the last half of the 1900s than in the first half, even though mean temperatures for February through May at Ithaca, New York, and Boston, Massachusetts, seemed to show only a very slight warming. The species advancing their arrivals included both seedeaters and insectivores. Thus, it is possible that many may actually have responded to significant temporal changes in the availability of various food types during the spring.

At the Long Point Bird Observatory on the north shore of Lake Ontario, two of five short-distance migrants for which adequate data were available showed significant advances in first spring arrival dates between 1975 and 2000. These species, the Ruby-crowned Kinglet and White-throated Sparrow, showed statistical advances that averaged 5.4 and 4.6 days per decade, respectively. When overall passage of birds during spring migration was considered, only the Brown Creeper showed significantly earlier migration, averaging 4.1 days per decade.

In southern Wisconsin, spring events, including the first singing of resident birds and first flowering of herbaceous plants and shrubs, were occurring about a week earlier in 1998 than in the 1930s. For arrivals of short-distance migrant birds, however, the evidence was equivocal. Only two of eleven species, the Red-winged Blackbird and the American Robin, showed significant trends toward earlier arrival. Both species tended to arrive about 10 days earlier in 1998 than in the mid-1930s.

Farther west, at Delta Marsh in Manitoba, Canada, a few short-distance migrants, such as the Mourning Dove, American Robin, Brown Creeper, Common Grackle, and Purple Finch, showed significant tendencies to arrive earlier in spring over the period 1939–2001. Many others, however, did not, but several of these did show significant tendencies to arrive earlier in years with warmer spring conditions.

Some altitudinal migrants are also responding to changing climate. American Robins returned to the Rocky Mountain Biological Laboratory in the Colorado mountains about 14 days earlier in spring in 1999 than in 1981. This pattern was shown even though the date of melting of winter snow had not changed.

## Spring Migration: Europe

Although advances in migration dates in North America are somewhat variable, in Europe short-distance migrants are clearly arriving substantially earlier in most localities than they did 50 to 100 years ago. This pattern is shown by an analysis of trends in spring migration for forty-four localities throughout much of Europe, from Italy in the south to Norway in the north, and from England to western Russia, over the period 1960–2006. For 104 species of short-distance migrants, first-arrival dates tended to advance about 5.6 days per decade, whereas median migration dates of passage of migrating birds advanced about 2.1 days per decade. A tendency was also noted for median passage dates to advance more rapidly in the most recent years. The tendency for earlier arrival was also stronger at intermediate latitudes than in the far south or far north. A second study also noted that between 1990 and 2000, short-distance migrants advanced spring migration dates more than long-distance migrants. Multiple-brooded species showed a greater advance in spring arrivals on breeding areas than did single-brooded species.

In the United Kingdom, other analyses show that short-distance migrants, such as the Common Chiffchaff, have mostly advanced their spring migration schedules. In Scotland, for example, most partial and short-distance migrants showed an advance in arrival date between 1974 and 2000.

In eastern Europe, observers have documented earlier arrivals by short-distance migrants at several locations. In Poland, between 1913 and 1996, seven species of short-distance migrants had advanced their spring arrival dates by about 20 days. Much of this change occurred during the period from 1970 to 1996. The species involved were as ecologically diverse as the Lapwing, Common Wood-Pigeon, Black Redstart, and European Serin. For all seven species considered together, correlations of arrival date with mean temperatures during February, March, and April were significant, with warmer temperatures linked to earlier arrival. In Lithuania, where winter and spring weather has warmed 1.5–2.0°C during the past century, several short-distance migrants have also significantly advanced their spring arrival. The Common Wood-Pigeon arrived more than 3 days earlier during the period 1990–2000 than during the period 1971–1987, and the Reed Bunting, more than 12 days earlier. On the Courland Spit in the eastern Baltic Sea, eleven of fourteen short-distance migrants advanced their arrival dates over a 44-year period. For one of these, the Song Thrush, the advance in arrival seemed to be favored by warmer conditions in its Iberian wintering area and increased frequency of favorable tailwinds during spring flights toward the Baltic region. Three species showed significant correlations with warm, moist spring conditions related to the positive phase of the NAO, and two of these three also showed earlier arrivals correlated simply with warm spring weather in the Baltic region.

Patterns are somewhat different in the most northern areas of Europe and farther east in Russia. Near Tromsö, northern Norway, for example, over the period 1970–2000, short-distance migrants tended to arrive earlier in years with warmer spring temperatures. However, no progressive trend of earlier arrival over the years was evident. Few data on movements of short-distance migrants are available from farther east in Russia. In the southern Ural Mountains, however, species such as the Rook and

European Starling appear not to have advanced their arrivals over the period 1971–2005. Although arrivals were indeed influenced by spring temperatures in different years, no progressive warming trend was evident in that region.

## Breeding Range and Time: North America

One of the most important long-term data sets about distribution and abundance of breeding birds in North America is the North American Breeding Bird Survey (BBS). The BBS is a system of roadside counts taken annually at about 2900 locations in Canada and the United States. Each count consists of observations at fifty points located at 0.8-kilometer intervals along a 40-kilometer route. BBS counts have been carried out since 1966. Analysis of these data, collected by many volunteer observers, is complicated by many factors but gives a continuing picture of population trends.

Some range shifts are occurring for short-distance migrants in apparent response to climatic warming. BBS data for the eastern United States and southeastern Canada show that since the mid-1960s, small land birds in general have shifted the northern limits of their breeding range about 23.5 kilometers per decade. Only two short-distance migrants, the Great-tailed Grackle and the Blue-gray Gnatcatcher, have shown such northward expansions, however, while six species have shown significant southward retreats. Breeding bird atlas data for New York State indicate that the southern limits of breeding ranges of several short-distance migrants, including the Pine Siskin, Purple Finch, White-winged Crossbill, and Brown Creeper, have retreated northward over a 20-year period.

Modeling efforts using BBS data suggest that substantial range shifts by many short-distance migrants are likely in the future. In the central Great Plains, Cassin's Sparrow and Lark Bunting appear likely to shift their ranges northward. In the northeastern United States and adjacent Canada, bioclimatic models combined with BBS data also indicate that many range shifts are likely. For forty-one short-distance migrants, these models examined scenarios of both high and low future emissions of greenhouse gases. For high-emission scenarios, fourteen species were projected to decrease in range area, seven to increase in area, and twenty to maintain about the same area. Many northern species were predicted to see their ranges retreat northward, while many southern species were predicted to spread northward.

A few widely distributed short-distance migrants have shown earlier breeding in North America over the past 25 to 50 years, but considerable variation exists among species and locations. Nest-record data for 1950–2000 show that two species in particular are tending to breed earlier in recent decades. The Red-winged Blackbird has advanced its breeding date about 7.5 days over the 50-year period. For the Eastern Bluebird, an advance in breeding date of about 4 days was seen during the period 1976–2000. Spring temperatures have significantly warmed throughout the breeding range of these species. In contrast, the American Robin and the Song Sparrow show no tendency for earlier breeding. For these species, however, spring temperature data throughout their very extensive breeding ranges also showed no overall tendency for general warming over the 50-year period.

Another short-distance migrant, the Tree Swallow, has shown an advance of about 9 days in mean egg-laying date across the northern United States and Canada between 1952 and 1992. The advance tended to be somewhat greater in western than in eastern localities and at southern relative to northern locations. The tendency for earlier breeding was closely related to the increase in mean temperature during May. Although Tree Swallows that nest earlier in the year tend to have the largest clutch sizes, these earlier-nesting swallows did not show clutches larger than expected.

## Breeding Range and Time: Europe

Ranges of many short-distance and partial migrants have shifted northward in Europe. In Britain, short-distance migrants breeding in southern parts of the country have seen their northern range margins shift farther north. In Finland, between 1974–1979 and 1986–1989 the northern boundary of the breeding range shifted farther north for many birds with breeding ranges located in the southern part of the country. Of about thirty-one species of short-distance migratory passerines, pigeons, and doves, eighteen species showed northward expansions. On the other hand, the northern breeding margin of another nine species retreated southward, and four species showed essentially no change.

In western Europe, many species of birds, both resident and migrant, have begun nesting earlier in the season. In the United Kingdom, some twenty species have shown an advance in egg-laying date, ranging from 4 to 17 days, between 1971 and 1995. Short-distance migrants advancing their egg-laying dates included the Corn Bunting, Chaffinch, and European

Greenfinch. For some of these species, such as the European Greenfinch, the trend toward earlier breeding seemed to be almost entirely a response to patterns of temperature and rainfall during March and April. Other species, such as the Chaffinch, showed a strong relationship between earlier egg-laying date and warmer temperatures but also a significant trend for earlier egg laying over the 56-year period, independent of temperature conditions. Populations of the European Starling also breed earlier in warmer years, the relationship being an advance of about 2 days per degree Celsius increase in mean temperature. On the Courland Spit in the eastern Baltic Sea, several species have also advanced their nesting dates.

In the northern and southern Alps of France, no consistent pattern of altitudinal shift was apparent in the ranges occupied by some twenty-four species of resident and migrant birds between 1973 and early 2000. Mean temperatures during spring and summer at both regions of the Alps had risen by about 2.3°C. Among short-distance migrants, most species showed no significant elevational shift. The European Blackbird, however, did show a significant upward shift of about 104 meters in the southern Alps, and the European Robin showed a shift of 114 meters in the northern Alps. On the other hand, the Firecrest showed a significant drop in elevational range of about 125 meters. It appears likely that changes in vegetational characteristics of the areas surveyed may be a complicating factor determining the altitudinal ranges utilized by these species.

# Populations

BBS data from 1966 through 2004 for the eastern and central United States and Canada show that short-distance migrants in general have experienced a 30 percent decline in abundance. The decline was most evident in species of open, edge, and wetland habitats. Short-distance migrants of forest habitats showed an increase in abundance of about 63 percent. Modeling studies that predict future frequencies of occurrence of forty-one species of short-distance migrants on BBS routes suggest that twenty to twenty-one species are likely to decrease, nine to increase, and eleven to twelve to remain about the same. An earlier analysis of BBS data from 1966 to 1988 indicated that many species breeding in early successional habitats in New England and wintering in similar habitats in the southeastern states had declined. These declines, however, were thought to relate to the disappearance of early successional habitats in the Southeast as biotic succession and commercial forestry converted more of the landscape to forests.

Several studies have documented the decline of migratory grassland birds in the central United States. These declines are clearly associated with the historical conversion of native grasslands to cropland and with more-recent intensification of agricultural practices. In 2009, the North American Bird Conservation Initiative noted that more than half of grassland birds have been declining. BBS data show that many migratory grassland birds of central North America have declined since the mid-1960s, particularly those with widespread and northern distributions (Table 6.2). These include the Loggerhead Shrike, Grasshopper Sparrow, Eastern and Western Meadowlarks, Lark Bunting, Baird's Sparrow, Lark Sparrow, and McCown's Longspur. The Field Sparrow, Dickcissel, Bobolink, Sprague's Pipit, and Scissor-tailed Flycatcher also suffered serious declines in the 1960s and 1970s. Change in precipitation pattern toward less frequent but more intense rainstorms, projected by some climate models, is likely to further stress these populations. In some areas, such as the Flint Hills of Kansas and Oklahoma, the long-term survival of some species is questionable.

In the western United States, patterns are less certain. Banding studies along the Rio Grande River in New Mexico indicate that over half of the short-distance migrants captured showed increases in numbers over the period 1985–1994. On the other hand, an analysis of BBS data for the western United States and for New Mexican birds with wide distributions across the United States have shown that some species of short-distance migrants showed significant increases in abundance, and about an equal number showed significant decreases. My personal experience with short-distance migrants in the Rio Grande Valley of northern New Mexico, where I live, is that multiyear cycles of drought and rainfall influence many of the species that pass through our area. Relating the population fluctuations of species such as the Bewick's Wren, Spotted Towhee, and Lark Sparrow to global climate is certainly a challenge. Nevertheless, this region experiences weather patterns related to ENSO and Pacific Decadal Oscillation cycles, which have changed in frequency, and the 2000–2003 drought that killed pinyons and Pondersoa Pines across much of our area is probably tied to global warming.

In the Pacific Northwest, reproductive success of eighteen species of short-distance migrants appears to be related to the NAO. Greater reproductive success occurs when the springtime NAO Index is positive, leading to cool weather that favors outbreaks of forest-defoliating insects. Several bird species that are specialist feeders on Western Spruce Budworm and Douglas-fir Tussock Moth larvae are highly benefited by these conditions.

TABLE 6.2. Population trends (annual % change) from breeding bird censuses for grassland birds of the central BBS region.

| | 1966–1979 | 1980–2006 | 1966–2006 |
|---|---|---|---|
| Widespread Plains Birds | | | |
| Greater Prairie Chicken[2] | 14.2 | –2.0 | –3.5 |
| Mountain Plover[2] | 3.7 | 2.8 | –3.1 |
| Loggerhead Shrike[1] | –4.7 | –2.7 | –3.3 |
| Bell's Vireo[2] | –10.1 | 1.0 | –3.3 |
| Horned Lark[2] | 36.2 | –2.3 | –1.8 |
| Brown Thrasher[2] | 0.0 | –1.9 | –1.1 |
| Brewer's Sparrow[1] | –2.5 | –3.0 | –3.6 |
| Field Sparrow[2] | –5.6 | –0.5 | –1.9 |
| Lark Bunting[1] | –3.3 | –1.9 | –1.5 |
| Lark Sparrow[2] | –6.0 | –0.8 | –2.4 |
| Grasshopper Sparrow[1] | –2.0 | –3.1 | –3.3 |
| Dickcissel[2] | –4.4 | 0.1 | 0.3 |
| Western Meadowlark[1] | –1.0 | –0.7 | –0.7 |
| Southern Plains Birds | | | |
| Scissor-tailed Flycatcher[2] | –4.0 | –0.4 | –0.3 |
| Cassin's Sparrow[1] | 0.4 | –0.9 | –1.7 |
| Henslow's Sparrow[3] | – | 9.1 | 11.2 |
| Painted Bunting[1] | –2.4 | 1.7 | –1.9 |
| Eastern Meadowlark[2] | 1.0 | –3.0 | –2.3 |
| Northern Plains Birds | | | |
| Sprague's Pipit[2] | –6.0 | –2.7 | –1.7 |
| Clay-colored Sparrow[1] | 8.4 | 1.3 | –0.2 |
| Vesper Sparrow[1] | 0.3 | –0.8 | –1.0 |
| Savannah Sparrow[1] | –0.9 | 1.6 | 1.2 |
| Baird's Sparrow[1] | –2.9 | –6.9 | –5.2 |
| LeConte's Sparrow[1] | 7.5 | 2.5 | 1.0 |
| Nelson's Sharp-tailed Sparrow[3] | – | 8.4 | 6.6 |
| McCown's Longspur[1] | 1.7 | 1.8 | 2.9 |
| Chestnut-collared Longspur[2] | 2.6 | –3.8 | –2.6 |
| Bobolink[2] | –3.4 | 0.7 | 0.1 |

Source: Sauer, J. R., J. E. Hines, and J. Fallon. 2008. *The North American Breeding Bird Survey, Results and Analysis 1966–2007*, Version 5.15.2008. USGS Patuxent Wildlife Research Center, Laurel, MD.
[1]Data reasonably good
[2]Data with some deficiency
[3]Data with significant deficiency

These birds include the Mountain Chickadee, Red-breasted Nuthatch, Ruby-crowned Kinglet, and Pine Siskin. These short-distance migrants tend to arrive on their breeding grounds earlier than long-distance migrants and thus appear to be better able to take advantage of the food provided by early-instar moth larvae.

## Fall Migration

Lengthening of warm summer weather in temperate and subarctic regions has extended the time that many short-distance migrants remain on the breeding range in both North America and Europe. Many short-distance migrants have thus shown a tendency toward later southward migration. At Long Point Bird Observatory on Lake Ontario, for example, all five short-distance migrants for which adequate data were available tended to show later fall migration between 1975 and 2000. For the Brown Creeper and Yellow-rumped Warbler the delays were statistically significant and averaged 5.8 and 3.5 days, respectively, per decade.

For fall migration in the Swiss Alps, a comparison was made of the dates of peak migratory passage of sixty-five species of birds for three periods: 1958–1969, 1970–1982, and 1988–1999. This analysis showed that twenty-eight of forty species of short-distance or partial migrant species had delayed their migrations to at least some degree. Most finches and thrushes had delayed their peak migration times, whereas sylviid warblers and regulid kinglets had advanced their migrations. Three other short-distance species that advanced their migrations were irruptive forms, whose movements are largely determined by mast abundance. The later migration time for at least some of these short-distance migrants may reflect an attempt at multiple broods by an increased fraction of their population.

Farther east in Europe, however, most short-distance migrants in the eastern Baltic region did not delay their southward migration. Only the Eurasian Blue Tit and the Eurasian Blackbird showed a tendency to migrate later.

## Wintering Patterns

In North America, the wintering areas of many short-distance migrants are shifting northward. An analysis of Christmas Bird Count data from 1975 through 2004 revealed that both resident and migratory species showed northward shifts in northern range boundaries and latitudinal centers of abundance during winter. On average, the northward shift in northern range boundaries over the 30-year period was about 44 kilometers, although it varied substantially among species. Species with northern range boundaries located in more southerly parts of North America tended to show greater northward shifts. A second analysis of Christmas Bird Count data, in this case from 1966–1967 through 2005–2006, showed that for migratory species with a midlatitude range above 35° N, forty-four showed northward shifts averaging 125.1 kilometers and nine showed southward shifts averaging 91.5 kilometers. For migratory species with a midlatitude range below 35° N, twenty-nine showed northward shifts averaging 46.9 kilometers and eleven showed southward shifts averaging 51.4 kilometers.

In New York and Massachusetts, in localities where a number of species of short-distance migrants did not overwinter in the first half of the twentieth century, wintering birds of several species have recently become common. These include the Eastern Bluebird, Yellow-rumped Warbler, White-throated Sparrow, Field Sparrow, and Brown-headed Cowbird.

One of these localities is Cape Cod, Massachusetts. Analysis of Christmas Bird Counts covering the period 1930–2001 revealed a substantial increase in number of wintering bird species with southern wintering affinities. In the 1930s, about thirty species of land birds with southern affinities were recorded, whereas over twice that number were noted in 2001. The ratio of species with southern wintering affinities to those with northern wintering tendencies increased from about 2.5 to nearly 5. The increase in ratio of southern to northern wintering species was particularly high during the decade of the 1990s, when regional warming was greatest. The ratios of southern to northern wintering species were positively correlated both with minimum winter temperature on Cape Cod and with an index of overall global warming. The changes in the wintering bird assemblage involved species of all major habitat types and were more closely related to weather changes than to changes in the extent of various habitats due to urbanization and other human activities.

In the northern Great Plains area of the midwestern United States, some short-distance migrants regarded as being limited in their northward distribution in winter by temperature are very responsive to year-to-year variation in coldness. For the Dark-eyed Junco, for example, the areas of greatest abundance in warm winters were located hundreds of kilometers north of areas of comparable abundance in cold winters. These results indicate that short-distance migrants can respond immediately to a changing temperature regime.

In eastern Ontario, Canada, conditions on the winter range may be reducing numbers of breeding female Red-winged Blackbirds, which have declined substantially over the period 1974–1995. This decline is evident in a reduction in the harem size of territorial males. The decline corresponds to the prolonged positive phase of the NAO. Increased female mortality over winter seemed to be the underlying cause of the reduced number of breeding females, possibly because more frequent and severe winter storms were affecting wintering blackbird populations in the southern United States.

Climatic warming is likely responsible for changes in the winter range of the Rufous Hummingbird. Since the 1970s, this species has expanded its winter range to include much of the southeastern United States north to Tennessee, southern Alabama, central Georgia, and the Carolinas. The traditional wintering range of this species is in northwestern and central Mexico. Many hummingbirds have been banded in the southeastern states, and recoveries of banded birds in subsequent winters indicate that many individuals have established a regular pattern of wintering in this region. They appear first as postbreeding birds in coastal areas of the Gulf States in July and August, much as postbreeding birds do in the interior western United States. Later, in November and December, they move to localities where they remain for the winter, so the immediate coastal region seems to be at least partly a migration corridor.

Patterns of wintering by short-distance migrants are changing in Europe, as well. Species such as the Eurasian Blackbird, a partial migrant in much of central Europe, are becoming more strongly resident in southern Germany, Switzerland, and France. For Redwings, there is also found a reduced tendency for birds from Finland to migrate as far south as France.

In Essex, UK, some wintering birds began arriving earlier or later and departing earlier or later over the period from 1966 to 2001. For most passerines, the differences are not statistically significant, but for the Snow Bunting, which arrives later and departs earlier to its Arctic nesting grounds, the duration of winter stay is significantly shorter.

As we shall see in Chapter 17, the Blackcap, an Old World warbler, has established a wintering population in England and Ireland. The birds of this population, coming from central Europe, follow a markedly different migration route than other Blackcaps and are apparently responding to the warming winter climate in the southern British Isles.

# Summary

Many studies in western Europe and several in North America suggest that short-distance migrants are arriving on their breeding grounds from several days to about 2 weeks earlier than 25–100 years ago. These responses appear stronger in Europe than in North America. Breeding ranges for many species have also shifted northward in parts of Europe, as have a few species in North America. Initiation of breeding has also advanced in both North America and Europe. Winter ranges of some species have also shifted northward. These patterns are directly attributable to climatic warming, especially at higher North Temperate latitudes. Weather patterns related to the NAO also appear to affect short-distance migrants in diverse fashion, encouraging earlier northward movements in spring, reducing survival of some species in stormy winter weather, and promoting abundant food supplies for others early in the breeding season.

## KEY REFERENCES

Butcher, G. S. and D. K. Niven. 2007. *Combining Data from the Christmas Bird Count and the Breeding Bird Survey to Determine the Continental Status and Trends of North American Birds*. National Audubon Society, Ivyland, PA.

Lemoine, N., H.-C. Schaefer, and K. Böhning-Gaese. 2007. "Species richness of migratory birds is influenced by global climate change." *Global Ecology and Biogeography* 16:55–64.

MacMynowski, D. P. and T. L. Root. 2007. "Climate and the complexity of migratory phenology: Sexes, migratory distance, and arrival distributions." *International Journal of Biometeorology* 51:361–373.

Miller-Rushing, A. J., T. L. Lloyd-Evans, R. B. Primack, and P. Satzinger. 2008. "Bird migration times, climate change, and changing population sizes." *Global Climate Change* 14:1959–1972.

Møller, A. P., D. Rubolini, and E. Lehikoinen. 2008. "Populations of migratory bird species that did not show a phenological response to climate change are declining." *Proceedings of the National Academy of Sciences USA* 105:16195–16200.

North American Bird Conservation Initiative. 2009. *The State of the Birds, United States of America, 2009*. U.S. Department of Interior, Washington, DC. 36 pp.

Rodenhouse, N. L., S. N. Mathews, K. P. McFarland, J. D. Lambert, L. R. Iverson, A. Prasad, T. S. Sillet, and R. T. Holmes. 2008. "Potential effects of climate change on birds of the Northeast." *Mitigating Adaptive Strategies for Global Change* 13:517–540.

Rubolini, D., A. P. Moller, K. Raino, and E. Lehikoinen. 2007. "Intraspecific consistency and geographic variability in temporal trends of spring migration phenology among European bird species." *Climate Research* 35:135–146.

Sauer, J. R., J. E. Hines, and J. Fallon. 2008. The North American Breeding Bird Survey, results and analysis 1966–2007. Version 5.15.2008. USGS Patuxent Wildlife Research Center, Laurel, MD. http://www.pwrc.usgs.gov/.

# Chapter 7

# High-latitude Land Birds: Nearctic–Neotropical Migrants

Nearctic–Neotropical migrants include species that breed in temperate to Arctic regions and winter in southern Mexico, the West Indies, Central America, and South America. In eastern North America, these species constitute about two-thirds of the breeding birds of deciduous and coniferous forests. Although the geographic barriers to migration are not as formidable as those between western Eurasia and Africa, many of these migrants must cross ocean areas such as the Caribbean Sea, the Gulf of Mexico, and the Gulf of California, as well as the deserts of northern Mexico and the southwestern United States.

For these migrants, climatic change is influencing conditions in breeding and nonbreeding areas and in the areas traversed in migration, as well as altering the major weather systems that influence the timing and efficiency of migratory movements. Understanding these influences is difficult because of the complicating influences of long-term climatic cycles and changing patterns of vegetation cover resulting from human activities.

## Spring Migration

Long-distance Neotropical migrants appear to have advanced their spring arrivals in North America only slightly, generally much less than the advances shown by short-distance migrants. This pattern exists in spite of the significant spring warming noted at most localities. Several major study sites have accumulated data on spring migration patterns, some relying on overall timing of migratory movements, others on first-arrival dates.

At the Manomet Center for Conservation Sciences on Cape Cod Bay, Massachusetts, data on the overall timing of spring migration cover the period 1970–2002. Here, as we noted earlier, spring temperatures increased by 1.5°C over these years. Of twenty-one Neotropical migrants for which good data were available, however, only five species showed significant advances in mean migration dates, and none showed an advance in first-arrival date. Advances for the five species averaged only slightly over 1 day per decade. Some Neotropical migrants also showed tendencies to show earlier mean migration dates in warmer springs or during the La Niña phase of the El Niño Southern Oscillation (see Chapter 2), when relatively warm conditions prevail in the southeastern United States. Overall, however, there was little evidence for substantial advance in migration dates for Neotropical migrants.

For many of the species at Manomet, including eleven of the twenty-one Neotropical migrants, the numbers of birds encountered in spring decreased over the years, a pattern that would tend, on average, to reduce the chance of detecting early appearance of migrating individuals. For one of these species, the Blackpoll Warbler, for example, no change was noted in mean passage date, but a significantly later first arrival was noted. The numbers of this species declined by two-thirds during the study period.

Detailed observations of spring migration cover a 40-year period at two other localities in eastern North America: Powdermill Nature Reserve in Pennsylvania, USA, and Long Point Bird Observatory in Ontario, Canada, on the north shore of Lake Erie. Mean temperatures for April and May in the general region of these observatories show only a slight trend toward warming—indeed, one not statistically significant. Considerable year-to-year variation exists, however. At the two observatories, most species of long-distance migrants showed trends toward earlier median capture dates over the years, but advances by only four species at Powdermill and none at Long Point were statistically significant. On the other hand, the fraction of species at Powdermill showing earlier mean capture dates over time rather than later, twenty-seven of thirty-two total, was highly significant. Still, the mean advance in mean capture date was only about 2 days.

Long-distance migrants reaching the Pennsylvania and Ontario observatories, however, were quite obviously responding to weather conditions in individual years. Mean capture dates at both localities were strongly related to mean spring temperatures. For each 1°C increase in mean spring temperature, captures tended to be 1 day earlier. In addition, the average number of days between mean capture dates along the Louisiana coast and those in Pennsylvania and Ontario were related to spring temperatures in eastern North America. For each 1°C rise in mean temperature, the time for migration between the Louisiana coast and the northern observatories was reduced by about 0.8 days. Thus, long-distance migrants were clearly responding to the temperature conditions they encountered by speeding their northward movement. These observations suggest that the earlier arrivals of at least some Neotropical migrants in eastern North America may be due simply to accelerated movements during later stages of migration.

Analysis of data on spring migration of long-distance migrants at Long Point Bird Observatory for the period 1975–2000 showed that differences were apparent when first arrivals were compared with total passage of migrants. First arrival dates for nine species of long-distance migrants showed that as a group they tended to arrive about 2.8 days earlier each decade. Analyzed individually, however, only the Yellow Warbler and the Common Yellowthroat showed significant tendencies for earlier first arrival. When all migration passage data were considered, a somewhat different pattern emerged. Data for the spring passage of each species were classified into five groups: earliest 5–35 percent, middle 35–65 percent, latest 65–95 percent, and the earliest and latest 5 percent. Analysis of the early, middle, and late 30 percent groups showed that only one long-distance migrant, the Least Flycatcher, showed a significant overall tendency for earlier migration.

The shores of Lake Erie, northern or southern, are some of the best areas in eastern North America to observe migratory birds. In early May, I have visited Crane Creek State Park and Magee Marsh Wildlife Refuge, east of Toledo, Ohio, on three occasions. The lake's south shore bills itself as the "Bird Watching Capital of Ohio." Birds on northward migration tend to stop on reaching the shoreline, and they pile up in the narrow, relatively open forest just back of the beach. At this time in May, mayflies often emerge in huge numbers, and the birds feed voraciously to rebuild their fat stores. Confined to this food-rich woodland, they tend to be oblivious to humans. My most recent visit was on May 7 and 8, 2009. On Sunday, International Migratory Bird Day, the park was jammed with birders, and the birds were there to delight them. The boardwalk through the swamp forest just inland from the shore was crowded with birders. I tallied twenty-four species of warblers in my 2-day visit, as well as many other long-distance migrants headed to Canada.

In Chicago, Illinois, data were collected for eleven species of long-distance migrants killed when they struck windows at McCormick Place from 1979 to 2002. In years with warmer spring temperatures, five of these species showed significantly earlier first arrival, beginning of main migration period, or median passage date. Eight species, however, showed correlations with the positive phase of the North Atlantic Oscillation (see Chapter 2), showing earlier dates for first arrivals, start of main migration period, or median passage date. This latter pattern is thought to reflect weather influences on departure from wintering areas and favorable conditions for movement along the southern portion of their migration routes.

Other studies have considered only first-arrival dates. The Maine study of migrant birds for the periods 1899–1911 and 1994–1997 (see Chapter 6) considered forty-three species of long-distance migrant land birds. Of these, twenty-eight showed no significant change in first arrival date. Four species showed a significant tendency to arrive 3 to 7 days earlier in the 1990s, but eleven species were significantly later in arrival, the delays varying from 4 to 10 days. At Middleborough, Massachusetts, three Neotropic migrants showed earlier arrivals over the period 1970–2002, but several others showed no advance in arrival.

Farther west, at Delta Marsh, in Manitoba, Canada, very few long-distance migrants exhibited earlier first arrivals over the period 1939–2001. Only the Common Yellowthroat, marginally a long-distance migrant, arrived significantly earlier. In addition, only one other long-distance migrant, the Baltimore Oriole, showed a significant tendency to arrive earlier in years with warmer springs.

At several locations in California, long-distance migrants show influences of both climatic warming and long-term climatic cycles. First arrival dates for twenty-one species were analyzed in relation to trends in spring temperatures and the climate patterns of the Pacific Decadal Oscillation (PDO), El Niño Southern Oscillation (ENSO), and North Atlantic Oscillation (NAO). During the period from 1969 to 2003, mean minimum spring temperatures tended to increase, but influences of ENSO and PDO phases also occurred. Of the twenty-one migrants considered, eight showed a trend toward earlier arrival and two toward later arrival. Overall, however, twelve species showed patterns of earlier arrival in years with warmer spring temperatures. Several species showed earlier arrival patterns correlated with positive ENSO and NAO indices. Positive ENSO phases (La Niñas) tend to promote cool, moist conditions on the California coast and southerly winds that might assist migrant flights. Exactly how positive NAO phases could influence these birds is uncertain, but they may improve conditions in wintering areas or create wind patterns that assist migrant flights.

## Breeding Range and Time

Data from the North American Breeding Bird Survey (BBS; see Chapter 6) show that eight species of long-distance migrants have pushed their northern limit of breeding farther north, whereas only two species show a southward retreat. The species spreading northward include the Blue-winged, Golden-winged, Hooded, and Kentucky Warblers; Black-billed Cuckoo; Summer Tanager; Swainson's Thrush; and Willow Flycatcher. The Alder Flycatcher shows a southward retreat of its northern breeding limit. These species are probably responding to climatic warming. The Cerulean Warbler has also shifted its breeding range northeastward by several hundred kilometers, although the underlying cause of this shift is not understood. Based on breeding bird atlas data for New York State, the southern limits of breeding ranges of several long-distance migrants, including the Bobolink, Least Flycatcher, Nashville Warbler, and Chestnut-sided Warbler, have shifted northward over a 20-year period. Most of these northward shifts almost certainly represent a response to climatic warming.

Modeling efforts, using BBS data, further suggest that in the future, major shifts are likely to occur in the distribution of Neotropic migrants in the northeastern United States and adjacent Canada. Using climate simulation models in combination with data on climate, elevation, the distribution of tree species, and the proportion of BBS routes with individual species, the responses of sixty-three Neotropical migrants were predicted. The models examined scenarios of both high and low future emissions of greenhouse gases. For high-emission scenarios, twenty-six to thirty species were projected to decrease the proportions of the BBS routes on which they were present, twenty-five species to increase their proportions, and eight to twelve to remain stable. The breeding ranges of fourteen to fifteen species were predicted to decrease in area, twenty-three to increase in area, and twenty-five to twenty-six to maintain about the same area. Many species would see their ranges retreat northward, while species typical of the southeastern United States would expand northward. These modeling analyses also predicted that greenhouse warming would restrict the location of high-quality territories of the Black-throated Blue Warbler to higher elevations and more northern localities. The Bicknell's Thrush, a high-elevation bird of spruce–fir forests, would likely lose most of its breeding habitat in the states of New York and Vermont.

Other modeling analyses also indicate that climate changes in North America during this coming century are likely to lead to major shifts in

TABLE 7.1. Estimated percentage losses of current breeding species of Neotropical migrant birds in the United States and gains of Neotropical migrant species by colonization from farther south based on climate change predicted by the Canadian Climate Centre's general circulation model for a doubling of atmospheric $CO_2$ level.

| United States Region | Species Lost | Percentage of Neotropical Migrant Birds | |
| --- | --- | --- | --- |
| | | Species Gained | Net Loss |
| New England | 44 | 29 | 15 |
| Mid-Atlantic | 45 | 22 | 23 |
| Southeast | 37 | 15 | 22 |
| Great Lakes | 53 | 24 | 29 |
| Eastern Midwest | 57 | 27 | 30 |
| Great Plains: Northern | 44 | 34 | 10 |
| Great Plains: Central | 44 | 36 | 8 |
| Great Plains: Southern | 32 | 18 | 14 |
| Rocky Mountains | 39 | 29 | 10 |
| Pacific Northwest | 32 | 16 | 16 |
| California | 29 | 23 | 6 |
| Southwest | 29 | 25 | 4 |

*Source:* Price and Root 2005.

ranges of many Neotropical migrants. Based on the Canadian Climate Centre for Climate Modelling and Analysis global circulation model, regions of the United States will likely lose from 29 to 57 percent of the Neotropical migrants that currently breed in them (Table 7.1). Because these losses will likely be compensated for by other Neotropical migrants that extend their ranges into these regions, net declines in number of breeding Neotropical species are estimated to range only from 4 to 30 percent.

Some observations in the midwestern United States suggest that failure to advance spring migration may cause long-distance migrants to miss the peak of food availability in their nesting areas. Most species of parulid wood warblers, especially those feeding primarily on forest canopy caterpillars, have not advanced their spring arrival dates at Fargo, North Dakota. Spring temperatures in North Dakota, however, reach conditions optimal for caterpillar populations 5.1 to 8.4 days earlier than in the early 1900s, whereas those in southern Illinois are 6.5 to 11.2 days later. If migrating warblers remain longer in Illinois because of the food availability there, the result might be that arrival on the breeding grounds in the northern United States and Canada will not occur until past the food peak at nesting locations.

## Population Dynamics

In eastern North America, concern has long centered on the population dynamics of long-distance migrants breeding in forests. In the Hutcheson Memorial Forest in New Jersey, for example, a substantial decline occurred in long-distance forest migrants between 1960 and 1984. Even in virgin mountain forests in West Virginia, six of fourteen Neotropical migrants disappeared between 1947 and 1983, with the total density of the remaining species being only 63 percent of that in 1947. In 1989, one of the first analyses of BBS data suggested that many long-distance forest migrants declined between 1978 and 1987, after having shown stable or increasing abundances from 1966 through 1977. Soon afterwards, analyses of BBS data for the period 1978–1988 showed, there were significant population declines for many wood warblers, thrushes, tanagers, flycatchers, and other forest birds in the eastern and central regions of North America.

Although some authors have raised doubts about declines of Neotropical migrant land birds, the evidence is now very strong that many species continue to decline substantially. The most recent BBS data show that trends for decline have continued for many Neotropical migrants in eastern and central North America. Neotropical migrant wood warblers that breed in eastern North America are among the species most affected (Table 7.2). Those with winter ranges that occupy large portions of Mexico, Central America, and northern South America experienced the greatest declines from 1980 through 2007. Some species that benefit from spruce budworm outbreaks, which have not occurred since the early 1970s, have shown especially large declines, presumably reflecting, at least in part, poor food availability on the breeding range. For others, such as the Cerulean Warbler, studies suggest that heavy adult mortality occurring on the winter range or in migration may be a contributor. Other Neotropical migrants of eastern North America that have also declined include several species of flycatchers, thrushes, and orioles, most of which winter in Central America and northern South America.

Other recent studies also suggest that declines of some long-distance migrants have occurred. Counts of migrating birds over 50 years in New England in the late 1900s, for example, show significant declines in eighteen of twenty-six species. The numbers of many species of Neotropical

TABLE 7.2. Population trends (annual % increase or decrease) for long-distance migrant Wood Warblers of eastern North America as estimated from North American Breeding Bird Survey data.

| Species | 1966–2007 | 1980–2007 | Wintering Area |
|---|---|---|---|
| Blue-winged Warbler | −1.1 | −1.8 | S Mexico to Panama |
| Golden-winged Warbler | −2.6 | −1.6 | S Mexico to Colombia |
| Tennessee Warbler[1] | −1.1 | −7.5 | S Mexico to NW S America |
| Orange-crowned Warbler | −1.1 | −1.2 | S USA to Guatemala |
| Nashville Warbler | 1.0 | 0.3 | Mexico to Guatemala |
| Northern Parula | 0.6 | 0.6 | Mexico to Honduras, West Indies |
| Yellow Warbler | 0.0 | −0.3 | Mexico to N S America |
| Chestnut-sided Warbler | −0.7 | −0.5 | S Mexico to Panama |
| Magnolia Warbler | 1.3 | 0.6 | Mexico to Costa Rica, West Indies |
| Cape May Warbler[1] | 0.5 | −2.9 | West Indies |
| Black-throated Blue Warbler | 0.9 | 1.6 | West Indies |
| Yellow-rumped Warbler | 0.1 | −0.5 | S USA to Panama, West Indies |
| Black-throated Green Warbler | 0.6 | 1.2 | Mexico to Panama, West Indies |
| Blackburnian Warbler | 0.7 | 1.0 | Panama to Peru |
| Yellow-throated Warbler | 1.0 | 0.8 | S USA to Panama, West Indies |
| Prairie Warbler | −2.0 | −1.5 | Florida to West Indies, Yucatán |
| Palm Warbler | 3.0 | 1.0 | S USA to Honduras, West Indies |
| Bay-breasted Warbler[1] | −1.9 | −3.5 | Costa Rica to Colombia |
| Blackpoll Warbler[2] | −9.7 | −9.5 | N S America |
| Cerulean Warbler | −4.1 | −2.9 | Colombia to Peru |
| Black and White Warbler | −0.8 | −1.3 | S USA to N South America, West Indies |
| American Redstart | −0.6 | −1.0 | S Mexico to N S America, West Indies |
| Prothonotary Warbler | −1.1 | −1.8 | Mexico to N S America, West Indies |
| Worm-eating Warbler | −1.6 | −3.6 | Mexico to Panama, West Indies |
| Swainson's Warbler | 4.5 | 3.3 | Mexico to Panama, West Indies |

Table 7.2. Continued

| Species | 1966–2007 | 1980–2007 | Wintering Area |
|---|---|---|---|
| Ovenbird | 0.3 | –0.1 | Florida to Panama, West Indies |
| Northern Waterthrush | –0.7 | –1.0 | S USA to N S America, West Indies |
| Louisiana Waterthrush | 0.7 | 0.3 | Mexico to N S America, West Indies |
| Kentucky Warbler | –0.9 | –0.8 | Mexico to Panama |
| Connecticut Warbler | –2.9 | –2.6 | N S America |
| Mourning Warbler | –1.5 | –2.5 | Panama to N S America |
| Common Yellowthroat | –0.5 | –0.8 | S USA to N S America, West Indies |
| Hooded Warbler | 1.9 | 1.9 | Mexico to N S America, West Indies |
| Wilson's Warbler | –2.3 | –2.9 | Mexico to Panama |
| Canada Warbler | –2.3 | –3.8 | N S America |

[1]Spruce budworm specialists
[2]Possible spruce budworm specialist

migrants encountered at the Manomet Center in Massachusetts in spring have also decreased over the years. At the Thunder Cape Bird Observatory, Ontario, on the north shore of Lake Superior, thirteen species of long-distance migrants showed significant trends of decrease in combined spring and fall numbers between 1995 and 2002, with only one species showing an increasing trend. Four additional long-distance migrants showed trends of decreasing numbers during fall migration.

Still other evidence for a substantial decline of migrants wintering in the Neotropics comes from radar observations of spring migrants arriving on the northern coast of the Gulf of Mexico. Flocks of migrants that depart from the Yucatán Peninsula early in the night on one day reach the northern coastline late in the afternoon of the next day. Arriving flocks are evident as masses of "blips" on weather surveillance radar screens. In the early 1960s, major flights were noted on 95 percent of days between early April and mid-May. From 1987 to 1989, however, major flights were noted on only 44 percent of all days, suggesting an overall transgulf migration only half as extensive.

In western North America, fewer studies of population trends are available. The BBS surveys from 1966 through 1988 found that most significant changes for forest birds were toward population increases, with some fly-catchers showing declines. From 1980 through 2007, declines were noted for several hummingbirds, flycatchers, and warblers, but changes were in most cases less than for birds in eastern and central regions. Some recent studies, however, suggest an overall decline of Neotropical migrants. Fall banding data along the Rio Grande River in New Mexico from 1985 through 1994 show declining trends for many species of Neotropical migrants. Substantial declines were noted for Wilson's Warbler and the West-ern Tanager.

In central California, trends in fall populations of thirty resident and Neotropical migrant birds at the Point Reyes Bird Observatory (PRBO) from 1979 to 1999 showed declines for six Neotropical migrants and eight residents; two Neotropical migrants and fourteen residents showed no sig-nificant trends. For Neotropical migrants the declines were greater during 1989–1999 than during the earlier decade. Trends for PRBO data for Neotropical migrants and BBS data for the same years for California and the Pacific Northwest did not show close correspondence, however. Thus, how closely the PRBO pattern reflects overall trends of western migrants is uncertain.

Changes in the periodicity of the ENSO appear certain to have impacts on Neotropical migrant songbirds. As we noted in Chapter 2, the interval between El Niños has decreased from nearly 6 years to about 2.2 years, and La Niña events are weaker. In Neotropical areas where many long-distance migrants winter, El Niños tend to favor dry conditions during the winter-ing period, while La Niñas favor wet conditions during this period. In North American breeding areas, the influences of El Niños and La Niñas vary depending on geographical location. In the Pacific Northwest, for ex-ample, El Niños tend to promote dry, warm summer conditions.

In New Hampshire, Black-throated Blue Warblers achieved greatest re-productive success in La Niña years, when the caterpillars that they feed on heavily were most abundant. Annual survival of individuals of this species wintering in Jamaica was also highest in La Niña years.

Yellow Warblers nesting in southern Manitoba also realized greatest nesting success and annual survival in La Niña years. In El Niño years, ear-lier spring warming leads to peaks of insect abundance that are earlier than optimal for feeding of young by nesting Yellow Warblers. Midsummer and fall conditions in this region are also typically wetter in El Niño years. Thus, suboptimal conditions for feeding young may underlie reduced nesting success in El Niño years. Exactly why annual survival should be highest in La Niña years is not entirely clear. Thus, for both Black-throated Blue and

Yellow Warblers, the increased frequency of El Niños, other things being equal, should tend to cause declines in their populations.

In the Pacific Northwest region, ENSO patterns were related to reproductive success of sixteen species of Neotropical migrant songbirds. Reproductive indices for these species were positively correlated with the index of ENSO precipitation for March through May. These effects seem to reflect the influences of rainfall, in increasing insect food availability in wintering areas in Mexico, and of favorable southerly winds during the period of northward migration. For some species, the contribution of ENSO conditions to outbreaks of Western Spruce Budworm and Douglas Fir Tussock Moth also favored reproductive success. These species included the Yellow Warbler, MacGillivray's Warbler, Wilson's Warbler, and Western Tanager.

Data from the BBS, combined with field studies in Pennsylvania, showed that populations of the Yellow-billed Cuckoo were strongly influenced by ENSO and NAO. In the case of ENSO, cuckoo populations tended to decline in the north-central United States after warmer, wetter El Niño years. The positive phase of NAO, on the other hand, showed a beneficial influence on cuckoo populations in the southern states, perhaps by promoting warmer winter conditions that favor large populations of caterpillars on which the species feeds. This species was one of the most rapidly declining Neotropical migrants over the BBS period from 1966 to 2002.

The increased frequency of hurricanes and severe tropical storms may also affect populations of land birds that migrate over the Atlantic Ocean, Gulf of Mexico, and Caribbean Sea. Analysis of BBS data suggests that population declines have occurred in some Neotropical songbirds that migrate over water from eastern North America to the West Indies and South America. In some species that have breeding ranges spanning eastern and western portions of North America, the declines are seen only in eastern populations. These include the Veery, Gray-cheeked Thrush, Swainson's Thrush, Mourning Warbler, and Rose-breasted Grosbeak. For these species, BBS indices were negatively correlated with the number of days with severe Atlantic storms during the preceding fall.

## Fall Migration

Patterns of change in fall migration in response to an extended summer period might be expected to vary with species, depending on several factors. Lengthening of the summer period might allow some species to increase the frequency of second nesting efforts or extend the period of postnuptial molt, thus delaying their fall migration. Alternatively, earlier completion of a single nesting might favor earlier return to the nonbreeding range if this confers survival advantages, such as early establishment of a nonbreeding territory.

At Long Point Bird Observatory in Ontario, Canada, for the period 1975–2000, only one of the eight species of long-distance migrants, the Blackpoll Warbler, showed a significantly later fall migration, whereas two species, the Yellow Warbler and Common Yellowthroat, showed significantly earlier migration patterns. These species also showed earlier first arrivals of individuals, probably mostly males, in spring. In contrast, the Least Flycatcher, although showing a significantly earlier spring migration pattern and also being single brooded, did not show earlier fall migration.

## Wintering Patterns

Climatic changes in the wintering areas of Neotropical migrants are certain to affect these species. Little information is yet available, however, on such effects. For American Redstarts wintering in Jamaica, weather conditions affect food availability, premigratory fattening, and time of departure on northward migration. Rainfall during the period from January through March determines the availability of insect foods and the condition of birds preparing for migration. The greater their body mass, the earlier redstarts depart on northward migrations. From 1995 through 2005, dry season rainfall in Jamaica declined substantially, and the departure dates of birds tended to become later. Future declines in rainfall in tropical wintering areas of North American migrants might thus alter the timing of spring migration, as well as reducing overwinter survival.

## Summary

Neotropical migrants have tended to advance their spring arrivals in North America in response to warmer spring weather much less than short-distance migrants. Climatic cycles of ENSO, PDO, and NAO also influence migration schedules. The breeding ranges of some Neotropical migrants are shifting northward, and substantial future shifts are predicted. Several studies have shown that many Neotropical migrants of forested habitats in eastern and central North America have declined, beginning in the 1970s and 1980s. Weather patterns related to ENSO and NAO apparently favor populations of some species and are detrimental to others. Some long-distance migrants are delaying southward migration in the fall, others advancing their departures. Weather conditions on wintering ranges also affect survival and timing of northward migration.

## KEY REFERENCES

Ballard, G., G. R. Geupel, N. Nur, and T. Gardali. 2003. "Long-term declines and decadal patterns in population trends of songbirds in western North America, 1979–1999." *The Condor* 105:737–755.

Gauthreaux, S. A. 1992. The use of weather radar to monitor long-term patterns of trans-Gulf migration in spring. Pp. 96–100 in J. M. Hagan III and D. W. Johnston (Eds.), *Ecology and Conservation of Neotropical Migrant Landbirds*. Smithsonian Institution Press, Washington, DC.

Hitch, A. T. and P. L. Leberg. 2007. "Breeding distributions of North American bird species moving north as a result of climate change." *Conservation Biology* 21:534–539.

MacMynowski, D. P., T. L. Root, G. Ballard, and G. R. Geupel. 2007. "Changes in spring arrival of Nearctic–Neotropical migrants attributes to multiscalar climate." *Global Change Biology* 13:1–13.

Marra, P. P., C. M. Francis, R. S. Mulvihill, and F. R. Moore. 2005. "The influence of climate on the timing and rate of spring bird migration." *Oecologia* 142:307–315.

Mills, A. M. 2005. "Changes in the timing of spring and autumn migration in North American migrant passerines during a period of global warming." *Ibis* 147:259–269.

Murphy-Klassen, H. M., T. J. Underwood, S. G. Sealey, and A. A. Czyrnyj. 2005. "Long-term trends in spring arrival dates of migrant birds at Delta Marsh, Manitoba, in relation to climate change." *The Auk* 122:1130–1148.

Nott, M. P., D. S. DeSante, R. B. Siegel, and P. Pyle. 2002. "Influences of El Niño/Southern Oscillation and the North Atlantic Oscillation on avian productivity in forests of the Pacific Northwest of North America." *Global Ecology and Biogeography* 11:333–342.

Price, J. T. and T. L. Root. 2005. *Potential Impacts of Climate Change on Neotropical Migrants: Management Implications*. USDA Forest Service General Technical Report PSW-GTR-191. Pp. 1123–1128.

Sauer, J. R. and S. Droege. 1992. Geographic patterns in population trends of neotropical migrants in North America. Pp. 26–42 in J. M. Hagen III and D. W. Johnston (Eds.), *Ecology and Conservation of Neotropical Migrant Landbirds*. Smithsonian Institution Press, Washington, DC.

# Chapter 8

# High-latitude Species of Land Birds: Palearctic Long-distance Migrants

Long-distance migrants between the western Palearctic and Africa face special challenges: crossing the Sahara and navigating across or around the Mediterranean Sea. The ability of these species to adapt to climatic change in their breeding and wintering ranges, as well as along their migration routes, is crucial to their survival. Adjusting migration schedules to assure arrival in breeding areas at the optimal time for rearing young is perhaps the most critical factor.

Many migrants from the eastern Palearctic also face challenges in their crossing of deserts of the Middle East, mountains of the Himalayan region, and heavily populated areas of China and southeast Asia. More than most other long-distance migrants, Palearctic–African and Palearctic–southeast Asian migrants are faced with the special problem that climatic changes in wintering areas are not necessarily correlated with those in their breeding ranges. Thus, these long-distance migrants are all faced with the difficulties of adapting to climatic changes in their wintering areas, along their migratory routes, and in breeding areas.

## Palearctic–African Migration System

Long-distance migrants between Europe and sub-Saharan Africa have shown some of the most serious impacts of climatic change.

## Overall Population Trends

Palearctic–African long-distance migrant birds of many kinds have experienced major population declines that appear to be related to changing climate. The Pan-European Common Bird Monitoring project, covering all countries of western Europe, found that from 1990 to 2005, population trends for 80 percent of the forty long-distance migrants for which adequate data were available showed declines. Declines were most frequent among nonforest birds, but four of the six forest species also declined markedly. Over the longer period from 1980 to 2005, population declines were noted in 78 percent of long-distance migrants.

Other recent analyses have shown similar conclusions. Analysis of populations of European long-distance migrants conducted by the Royal Society for the Protection of Birds concluded that twenty-seven of thirty-seven species migrating to sub-Saharan Africa have declined in abundance. Over the period 1980–2005, the Common Nightingale, for example, has declined by 63 percent over its European range. Of thirty-six species of sub-Saharan migrants that have bred in Britain, two have disappeared, eleven have declined by more than 50 percent, and another seven species have declined significantly in abundance.

Still another recent analysis examined population trends for 359 species, including 121 species of long-distance migrants, over the period 1970–2000. Of these, 68 species were small nonraptorial land birds migrating between the western Palearctic region and sub-Saharan Africa. At least 21 species of these land birds showed significant patterns of decline during all or part of this period, whereas only 12 species showed patterns of increase. For 18 long-distance migrants paired with nonmigrant congeners breeding in the western Palearctic, the difference was striking, with declines of the migratory members being significantly greater. Only one long-distance migrant member of these pairs, the Citrine Wagtail, showed a significant increase throughout the 30-year period.

Local studies show similar patterns. In France, an analysis of population trends for long-distance migrant species over the period 1989–2005 showed twice as many declines as increases. There was a strong tendency for declines to occur in species with low tolerance of high temperatures, suggesting that the warming climate in the region may have contributed to their declines. Nevertheless, in 2003, a year with warmer than average spring temperatures, trans-Saharan migrants had greater than average reproductive success. In another localized study, long-distance migrants in the Lake Constance area of Germany, Switzerland, and Austria decreased in abundance by slightly over 46 percent between 1980 and 2002.

Decline of populations of many species began in the 1970s. European species wintering in Africa declined more than other migratory birds during the period 1970–1990. In addition, their decline during this period was greater than during 1990–2000, suggesting that the intense sub-Saharan drought during the former period was a substantial contributor to declines. Farmland birds, several of which are migrants to sub-Saharan Africa, also declined in populations during 1970–1990. The decline in populations since 1990 may have a different cause. Species that have not advanced their spring migration dates are those suffering the greatest recent population declines, which seems to signify that arrival of these species on breeding areas does not coincide with conditions optimal for reproductive success.

Clearly, general environmental changes are acting to the detriment of many migratory land birds in the Palearctic–African system. Many of the species winter or prepare for northward migration in the Sahel region of West Africa, where drought and agricultural intensification have damaged or destroyed the natural vegetation in recent decades. Climatic changes have occurred along migration routes north of the Sahara, as well as in breeding areas, altering the timing of migratory movements and arrival on breeding grounds. It is also possible that milder weather conditions in the western Palearctic breeding areas have played a role through competition between long-distance migrants and species that are residents or short-distance migrants.

## Spring Migration

The timing of long-distance migration has changed in many cases. Many migratory species breeding in the western Palearctic have tended to arrive earlier in spring during the latter part of the twentieth century and the early years of the twenty-first century. This tendency seems to have increased in recent decades. An analysis of first-arrival dates for 19 species of long-distance migrants in Leicestershire, England, over the period 1942–1991, for example, found only two species that showed significantly earlier arrival, the Bank Swallow and Sedge Warbler, but four that showed significantly later arrival: Cuckoo, Greater Whitethroat, Whinchat, and Garden Warbler. Many of these trends may reflect the warm spring conditions that were strongly correlated with early arrivals during the 1940s and early 1950s. Several long-distance migrants in Spain also showed a tendency for early spring arrivals during this same period.

In the British Isles and western Europe, the tendency for many long-distance migrants to arrive substantially earlier in spring since the 1960s is very strong. In an analysis of spring migration patterns for forty-four localities, covering the period 1960–2006, eighty species of long-distance migrants showed that first-arrival dates tended to advance about 2.2 days per decade, although median dates of passage of migrating birds advanced very little. In Oxfordshire, England, the first-arrival dates of seventeen of twenty species of long-distance, trans-Saharan migrants had advanced about 8 days during the period 1971–2000. The advances in arrival were strongly correlated with mean winter temperatures in sub-Saharan Africa, suggesting that their earlier arrival in England was due primarily to earlier departure from Africa. In Leicestershire, several additional long-distance migrants appear to show trends toward earlier arrival since the mid-1950s.

In Spain, some trans-Saharan migrants, including the Barn Swallow, Common Swift, and Cuckoo, have advanced their spring arrivals since the mid-1970s. No advance was detected for the Common Nightingale. All four species, however, showed a significant tendency to arrive earlier in warmer springs. At one site in northeastern Spain, in contrast, the first spring arrival dates of these same species appeared to have become later over the period 1952–2003. For three, the Barn Swallow, Common Nightingale, and Cuckoo, the pattern was clearly for delayed arrival. The Hoopoe showed a slight tendency for earlier arrival since about 1975. These trends also appear to be correlated with temperature and precipitation patterns in West Africa. Temperatures higher than normal in West Africa tend to delay arrival, and more rainfall than normal tends to accelerate arrival in Spain. Mild temperatures and high rainfall are thought to favor abundant foods that enable migratory birds to initiate migration earlier than average.

On the island of Helgoland, in the southeastern corner of the North Sea, the mean passage times of ten of twelve long-distance spring migrants advanced by 0.45–2.6 days per decade between 1960 and 2000. Nine trans-Saharan migrants seem to be arriving earlier both in southern Italy and in Scandinavia over the past 20–30 years. On the other hand, a recent analysis of spring migration dates and population trends found that during the period 1990–2000, long-distance migrants showed the least tendency to advance their spring migration dates. At Ottenby Bird Observatory on Öland Island, Sweden, eighteen of the nineteen species of long-distance migrants were tending to advance their mean passage time. On average, however, the change was slightly less than 1 day per decade. The advance in migration time appeared to be generally similar for species migrating along western routes and species following eastern routes through Europe from their wintering areas. This observation is somewhat puzzling, since the weather associated with positive North Atlantic Oscillation (see Chapter 2) in southern (Mediterranean) Europe is warm and moist in the west but dry and cool in the east.

Slightly farther south, migration dates for spring migrants over the period 1976–1997 on Christiansø Island in the Baltic Sea near Bornholm were summarized to show the first individual, first 5 percent, first 50 percent, and first 95 percent of individuals for each species. Data for all seventeen species of long-distance migrants showed a tendency for earlier arrival, ranging from 0.24 days per year for the first individual to 0.12 days per year for the first 50 percent of individuals. The tendency for earlier migration varied substantially among species, being least in several species of warblers and the Thrush Nightingale. Species with the longest migration distances also showed the least advance in migration timing. Furthermore, species with declining populations also showed the least advance in migration dates. No pronounced differences were evident for species wintering in west, central, or east Africa.

Farther east, advances have also tended to occur for many long-distance migrants. In western Poland, eight species of long-distance migrants showed a trend toward earlier arrival over the period 1913–1996. Most also showed a stronger trend for earlier arrival over the period 1970–1996. Over this latter period, however, two species, Lesser Whitethroat and Whinchat, showed a trend for later arrival. In Lithuania, twelve species of long-distance migrants arrived earlier during the period 1990–2000 than during 1971–1987. These species advanced their arrivals by 3.6 to 10.2 days. Only one species, the Wryneck, showed a significantly delayed arrival. The Lesser Whitethroat and Whinchat showed earlier arrival, in contrast to the situation reported in Poland. Many other long-distance migrants showed no significant change in arrival dates. Long-term studies at a banding station on the Courland Spit, in the eastern Baltic Sea, have shown that eight of eleven long-distance migrants have advanced their first or median arrival dates significantly. Most were also influenced to arrive earlier by warm temperatures in the Baltic region, as well as in Spain, where many are passage migrants.

Still farther east and north, the situation may be somewhat different. In the southern Ural Mountains, several long-distance migrants appear not to have advanced their arrivals over the period 1971–2005. Although arrivals were influenced by warmer or colder spring weather in different years, no progressive climatic warming trend was evident. In the far north, near Tromsö, northern Norway, migratory birds, including several long-distance migrants, have tended to arrive earlier in years with warmer spring temperatures. No progressive trend appears to exist, however, toward earlier arrival over the period 1970–2000. These patterns are similar to those shown by short-distance migrants.

Analyses focused on particular species have revealed more specific patterns. In Europe, for example, spring arrival of the Barn Swallow has advanced most at lower latitudes and least at the most northern localities. In Slovakia, spring arrival has occurred about 2.1 days earlier for each 1°C warming of spring temperatures, whereas in England it has advanced 1.6–1.8 days and in Finland only 1.2 days for the same warming.

For the Willow Warbler, spring arrival has tended to advance at five banding stations in Europe, the westernmost being in Kent, England, and the easternmost on Lake Ladoga, Russia. Mean spring passage of migrating birds trapped and banded has advanced most on the Kent coast, about 5 days per decade. At eastern stations, the first spring arrivals have tended to become earlier, whereas mean spring passage has advanced less.

The European Pied Flycatcher presents an interesting case, in which spring migration pattern apparently varies across Europe. Birds breeding in western and central Europe show earlier migration schedules than those breeding in northern and eastern Europe. In the Netherlands, and apparently in Spain and other western European areas, mean spring arrival dates of European Pied Flycatchers since 1969 have not shown an advance. Males, however, do arrive earlier in years with warmer temperatures at stopover sites in North Africa and at their nesting areas in the Netherlands. Farther east and north, however, the birds have advanced their spring arrival. Spring migration at Helgoland was about a week earlier in 2000 than in 1960, and the arrival of birds at Christiansø Island and the Courland Spit in the Baltic Sea and in southwestern Finland has become significantly earlier. Spring arrivals in these areas are also correlated with the North Atlantic Oscillation (NAO), tending to be earlier following warm, moist winters. Birds that have not advanced their arrivals are those that do not encounter warmer temperatures during migration, while those that have advanced their arrivals encounter more favorable temperatures and consequently speed their migratory journeys.

The trend toward earlier spring migration in the western Palearctic is also correlated with the North Atlantic Oscillation Index, which has tended to show a more frequent positive condition through the last half of the twentieth century. At Helgoland, the passage times of all twelve long-distance spring migrants were also significantly related to the NAO. The more positive the NAO, indicating warm, moist winter conditions, the earlier was the mean passage time for migrants. The earlier passage was most likely due to more favorable food availability that enabled birds to accelerate their migratory pattern. On Jurmo Island and the Hanko Peninsula of southern Finland, thirteen species of long-distance migrants also showed strong trends for earlier passage during positive phases of the NAO. At Ottenby, Sweden, a similar advance of mean passage time occurred during positive phases of the NAO for sixteen of nineteen long-distance migrants. On the Courland Spit, in the eastern Baltic Sea, arrivals of most long-distance migrants also showed significant relation to the NAO, being earlier during the positive phase of the oscillation.

Changes in the timing of spring migration apparently reflect different climatic influences. For the European Pied Flycatcher, and probably other species, earlier arrival on the breeding grounds seems to reflect faster migratory movement through areas north of the Sahara. Other species appear to benefit from climatic changes in sub-Saharan areas that favor earlier departures for the north, as well as faster movement through North Africa and southern Europe.

## Breeding Range, Time, Success

Breeding ranges of many long-distance migrants have shifted northward in Europe. In the British Isles, many land and freshwater birds with southerly distributions have moved north. In Finland, the northern breeding limit for many long-distance migrants shifted northward between 1974–1979 and 1986–1989. Eighteen of thirty species showed this northward expansion, although another ten species retreated southward, and two species showed essentially no change. Most of the northern extensions of range margins were for species previously limited to southern Finland.

Changing climate is also likely to alter the competitive relations among migrant species and between migrants and residents in many areas. In the Czech Republic, for example, the European Pied Flycatcher appears to be displacing the Collared Flycatcher under a climate reflecting the more frequent positive influences of the NAO. Warmer winters in the Lake Constance area on the border of Germany and Switzerland are correlated with the decline of breeding long-distance migrants relative to short-distance migrants and permanent residents.

Many European species are showing earlier breeding. In the United Kingdom, at least six species of long-distance migrants advanced their breeding dates between 1971 and 1995. For some species, such as the Willow Warbler and Spotted Flycatcher, the change was mostly attributable to patterns of temperature and precipitation in the breeding areas. For others, including several long-distance migrants, trends toward earlier breeding were independent of changing influences of temperature and rainfall. At a banding station on the Baltic Sea coast, several long-distance migrants, including the White Wagtail, European Pied Flycatcher, Barn Swallow, and Lesser Whitethroat, showed a trend toward breeding earlier over the period 1959–2002, as evidenced by earlier captures of young of the year. These and many other long-distance migrants also showed significant trends for breeding earlier in warmer spring and early summer conditions.

In southwestern Poland, where summer temperatures warmed substantially between 1970 and 2006, Eurasian Reed-Warblers have begun nesting much earlier (Figure 8.1). Initiation of nesting has advanced about 3 weeks, and the median date for laying of the first egg about 2 weeks. In addition, the end of the nesting season has remained nearly the same, so more pairs are now able to rear second broods. In the 1970s and 1980s,

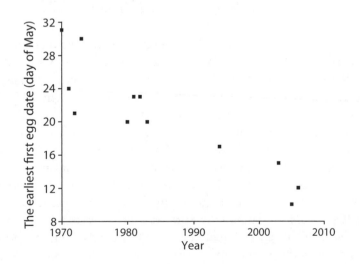

FIGURE 8.1. Temperature and initiation of nesting by Eurasian Reed-Warblers in southwestern Poland. Earliest first egg-laying date from 1970 through 2006. (Modified from Halupka et al. 2008.)

0 to 15 percent of pairs attempted second nestings, whereas in 1994–2006, 35 percent of pairs reared second broods. Nest predation has also declined, apparently because reed beds are showing earlier growth in spring, providing better concealment of nests in early summer.

The European Pied Flycatcher has shown a complex response to changing climate in Europe. In central Spain, European Pied Flycatchers did not advance their time of egg laying between 1984 and 2001, in spite of the fact that May temperatures had warmed and leafing out by oaks had begun earlier in the spring. In this region, leafing of oaks began about 8 days earlier over the period 1981–1991. General observations suggested that the peak of insect food abundance was also shifted to an earlier date, so the peak demand for food by nestlings became out of phase with food availability. Over the study period, the time of initiation of nesting and the mean clutch size of the birds did not change, but breeding success declined substantially. Breeding success was negatively correlated with mean May temperature — the warmer the weather, the poorer the success.

In contrast, in the Netherlands, although the European Pied Flycatcher has not tended to arrive earlier in spring, it advanced its mean egg-laying date by about 10 days in the period between 1980 and 2000. This advance, however, was not enough to match the seasonal advance in optimal food availability. This was reflected in the greater reproductive success of those flycatcher pairs nesting earliest. Further analyses of the laying dates of European Pied Flycatchers at localities across Europe showed that the advance in laying dates varied greatly and was greatest where spring temperatures had warmed most. A decline of about 90 percent in many populations of this species appears to be the consequence of the mismatch between nesting date and optimal prey availability date. Advances in laying date have occurred throughout European midlatitudes, most strongly in eastern areas. In northern Scandinavia, in contrast, even though faster migratory movements have led to earlier arrival, the laying date of European Pied Flycatchers became delayed over the period 1980–2004.

In Israel, several migratory species from tropical Africa have recently established breeding populations in southern parts of the country. These include two water birds, the Namaqua Dove, and the Black Scrub-Robin. In addition, the Eurasian Golden Oriole, a common passage migrant in Israel to breeding ranges farther north, has also established a small breeding population in northern Israel. A number of other migrants that breed in Israel have expanded their ranges. The fact that more species with southern than northern affinities have established or expanded breeding populations suggests that climatic warming is a significant influence.

So far, altitudinal shifts by breeding birds in response to warming spring and summer weather have received little attention. At two sites in the French Alps, where spring temperatures had warmed about 2.6°C, no consistent change was noted in elevational limits for four species of long-distance migrants between 1973 and 2000. Changing temperatures appeared to be less important than other factors, such as habitat suitability and interspecific interactions.

## Fall Migration

The response of long-distance migrants to extended summer conditions has varied considerably among species. In Oxfordshire, England, sixteen of the twenty species are showing earlier departures, averaging about 8 days. These earlier departures show a significant correlation with minimum summer temperatures, which have increased significantly during the 30-year period of the analysis. Similarly, the mean time of southward passage of twenty of twenty-five species advanced at Col de Bretolet, Switzerland, during the period 1958–1999. These migrants included the Willow Warbler, Garden Warbler, and European Pied Flycatcher. Most of these were single-brooded species. Species that tended to be double brooded, on the other hand, did not advance their fall migration dates. Earlier southward flights may be advantageous in enabling birds to cross the Sahel zone on the southern edge of the Sahara before the onset of the dry season in that region. At the Courland Spit in the eastern Baltic Sea, only two of nine species of long-distance migrants, the European Pied Flycatcher and the Lesser Whitethroat, tended to advance the dates of southward migration. Another western Palearctic species, the Common Swift, has also advanced its fall migration.

Some long-distance migrants have tended to delay their fall migration from the western Palearctic. At the Helgoland banding station, these include the Willow Warbler, Common Redstart, and Ring Ouzel. Thus, most of these species are spending more time on the breeding range during summer. Another six species showed no change in timing of fall migration.

## Staging and Stopover Relationships

How climate change will affect the stopover ecology of long-distance migrants is uncertain. Most western Palearctic species appear not to show high fidelity to particular stopover sites, except perhaps those that are near breeding or wintering areas to which they are faithful. Nevertheless, if changing climate reduces the suitability of stopover sites along migration routes, populations of these species could be affected detrimentally.

Stopover relationships during fall migration could be affected if climate change alters the phenology of food resources in stopover localities. Sedge Warblers and Great Reed-Warblers, for example, utilize reed-bed habitats in southern coastal France as migration stopover sites. Both species have advanced their spring arrivals in Europe. Fall migration, correlated with spring temperatures, has also become earlier. Neither species has extended fall stopover duration, suggesting that food availability in these sites has continued to be adequate and that there is an advantage to reaching wintering areas earlier.

Some passerines migrating from Europe to sub-Saharan Africa cross both the Mediterranean Sea and the Sahara in a single flight involving two nights and one and a half daylight periods. Other birds cross the Mediterranean Sea in a night flight and stop over during the following day on the northern coast of Africa. They then continue the next night on a flight across the Sahara, probably in most cases another single flight. Some birds, such as Western Yellow Wagtails, apparently do stop in the desert at oases, and it may be that some other small passerines do not make the Sahara crossing in a single nonstop flight. Nevertheless, the lack of food and water necessitates that they must store resources adequate for the entire crossing even if they make stops during the crossing. Staging areas along the northern coast of the Mediterranean and stopover areas in North Africa are thus probably essential to some trans-Saharan migrants.

Passerine birds crossing the Sahara in spring and fall would seem to be particularly at risk if the crossing distance were to increase because of drier conditions in North Africa and sub-Saharan Africa. For species like the Garden Warblers, which effectively double their body mass by fat deposition immediately prior to the crossing flight, an extension of the required flight distance could make the crossing impractical.

Birds returning north in spring presumably behave in a similar fashion. Garden Warblers apparently cross both the Sahara and the Mediterranean without stopping in North Africa to replenish fat deposits. Despite these remarkable abilities, the weather conditions of staging areas, such as the Sahel in spring, are obviously critical to successful migrations.

It appears likely that differences in arrival patterns of long-distance migratory birds in different parts of Europe are related, at least in part, to conditions at stopover sites in North Africa and southern Europe. Although

the European Pied Flycatcher has not advanced its spring migration schedule enough to maintain an optimal relation to food resources for young, it
has accelerated its migration pattern somewhat. Arrival of males at the
breeding areas in the Netherlands is influenced by warmer temperatures in
both the European breeding areas and in stopover areas in North Africa.
The climate in these North African sites, particularly in Morocco, has
warmed substantially between mid-March and mid-April. Barn Swallows
also arrive earlier in Denmark in years when dry, warm conditions in North
Africa apparently reduce the availability of insect foods.

### Winter Range

Little is known about the effects of changing conditions in sub-Saharan
winter ranges on long-distance migrants. In the Sahelian region of northern Nigeria, wintering Palearctic warblers are reported to be surviving well
in spite of habitat degradation. The Greater Whitethroat appears to be tol-
erant of even severe habitat degradation, as long as the shrub *Salvadora per-
sica* survives in abundance adequate to provide the fruit that this species depends on for premigratory fattening.

## High-latitude Species of Land Birds:
## Palearctic–Southeast Asian

Very little information is available on the response of migratory birds to climatic change in eastern Asia, although climatic change is substantial. In
Japan, for example, annual mean temperature has increased by about 1.0°C
during the twentieth century, with much of the warming reflecting milder
winter temperatures. There have been substantial changes in the occurrence
and abundance of many migratory birds, although the exact role of climatic
change is uncertain. A recent statistical analysis of range changes by breeding birds in Japan found that some long-distance migrants are declining. Ranges of the Gray Nightjar, Tiger Shrike, Brown Shrike, and Yellow-
breasted Bunting had shrunk by more than 50 percent. Climate change was
considered to be one possible contributor, along with habitat destruction.

Several other studies have shown declines in long-distance migrants in
specific parts of Japan. In central Honshu, for example, Japanese Paradise-
Flycatchers and Ashy Minivets have declined in abundance in forest areas.
Near Tokyo, surveys of breeding birds in the 1970s and 1990s revealed that
a dozen tropical migrants declined or disappeared from upland forest areas

where habitat conditions had not changed significantly. These included the Japanese Paradise-Flycatcher, Ashy Minivet, Gray Nightjar, Oriental and Lesser Cuckoos, Tiger Shrike, Siberian Blue Robin, Sakhalin Warbler, Eastern Crowned Leaf-Warbler, and Siberian and Scaly Thrushes. Several long-distance migrants also disappeared in southern Honshu between 1973 and 1995. Other species declining between 1972 and 1999 at one or more study areas on Honshu included the Japanese Thrush, Dollarbird, Yellow-breasted Bunting, and Ruddy Kingfisher. The Eurasian Skylark has also declined in central Japan, although habitat loss may be a major factor.

Migrants have also declined at several localities on the northern island of Hokkaido between 1972 and 1999. At Naporo, Hokkaido, for example, the White Wagtail, Eurasian Skylark, Brown-headed Thrush, Chestnut-eared Bunting, and White-cheeked Starling disappeared as breeding birds. Between 1991 and 2001, the Common Cuckoo and Yellow-breasted Bunting largely disappeared from eastern Hokkaido.

One of the few detailed analyses of changes in breeding schedules by migratory birds documented earlier nesting by the Chestnut-cheeked Starling, a species that breeds in Honshu and winters in Borneo. Over the period 1978–1998, the earliest egg-laying date advanced 14.6 days in central Honshu. In addition, clutch size has tended to increase. Over the 21-year period, the average clutch increased by about 1.2 eggs and is now about 6 eggs per nest attempt. Although there appeared to be no significant correlation between the advance of egg laying and maximum daily temperatures in Honshu, maximum daily temperatures in the winter range at Kota Kinabalu, Sabah, and along the migration route at Naha, Okinawa, had warmed significantly during the observation period.

Habitat fragmentation may compound the influence of climate change for some Japanese birds. A recent analysis of the relationship of Japanese songbirds to fragmented forest environments showed, for example, that five species of tropical migrants showed some degree of area sensitivity, tending to disappear from small areas of forest. Another eleven species did not show area sensitivity.

# Summary

Many trans-Saharan migrants have shown major population declines over the last 20–30 years. Many species also show earlier northward migration in spring, although the advances are much less than for short-distance migrants. Much of the advance in migration times appears to be related to faster movements along the northern portions of the migration routes.

Many species have extended their breeding ranges northward and advanced their breeding schedules. Some species have advanced their southward migration schedules in the fall; others have extended their stay on the breeding grounds. The little information available from eastern Asia suggests that breeding by some long-distance migrants has also become earlier in the season and that declines in some species have also occurred.

## KEY REFERENCES

Bairlein, F. and O. Hüppop. 2004. Migratory fuelling and global climate change. Pp. 33–47 in A. P. Møller, W. Fiedler, and P. Berthold (Eds.), *Birds and Climate Change: Advances in Ecological Research*. Vol. 35. Elsevier, Amsterdam.

Both, C., A. V. Artemyev, B. Blaauw, R. J. Cowie, A. J. Dekhuijzen, T. Eeva, A. Enemar, et al. 2004. "Large-scale geographical variation confirms that climate change causes birds to lay earlier." *Proceedings of the Royal Society of London B* 271:1657–1662.

Both, C., S. Bouwhuis, C. M. Lessells, and M. E. Visser. 2006. "Climate change and population declines in a long-distance migratory bird." *Nature* 441:81–83.

Both, C. and L. te. Marvelde. 2007. "Climate change and timing of avian breeding and migration throughout Europe." *Climate Research* 35:93–105.

Cotton, P. A. 2003. "Avian migration phenology and global climate change." *Proceedings of the National Academy of Sciences USA* 100:12219–12222.

Halupka, L., A. Dyrcz, and M. Borowiec. 2008. "Climate change affects breeding of reed warblers *Acrocephalus scirpaceus*." *Journal of Avian Biology* 39:95–100.

Jenni, L. and M. Kéry. 2003. "Timing of autumn bird migration under climate change: Advances in long-distance migrants, delays in short-distance migrants." *Proceedings of the Royal Society of London B* 270:1467–1471.

Jonzén, N., A. Lindén, T. Ergon, E. Knudsen, J. O. Vik, D. Rubolini, D. Piacentini, et al. 2006. "Rapid advance of spring arrival dates in long-distance migratory birds." *Science* 312:1959–1961.

Møller, A. P., D. Rubolini, and E. Lehikoinen. 2008. "Populations of migratory bird species that did not show a phenological response to climate change are declining." *Proceedings of the National Academy of Sciences USA* 105:16195–16200.

Sanderson, F. J., P. F. Donald, D. J. Pain, I. J. Burfield, and F. P. J. von Bommel. 2006. "Long-term population declines in Afro-Palearctic migrant birds." *Biological Conservation* 131:93–105.

Tøtterup, A. P., K. Thorup, and C. Rahbek. 2006. "Patterns of change in timing of spring migration in North European songbird populations." *Journal of Avian Biology* 37:84–92.

# Chapter 9

# *Land Birds of the Temperate Southern Hemisphere*

Climate change is affecting continental areas of the Southern Hemisphere, leading to change in migratory patterns of land birds in Australia, South America, and South Africa. The migrations of land birds in these areas are almost entirely confined to the same continental areas, as are their adjustments to changing climate. Awareness of changes in migration patterns has emerged only recently in these areas, and their documentation is difficult because of the shortage of long-term data sets that reveal trends.

South America, with a continental area extending much farther south than either Australia or South Africa, has the largest latitudinal migration system. Australia and South Africa have roughly half as many migratory land and freshwater birds. In addition, they exhibit strong influences of aridity, coupled with mild temperatures, that strongly influence patterns of seasonal movements. Both continents exhibit a substantial number of species that are nomadic, in the sense of showing irregular movements tied to large seasonal and interannual patterns of change in rainfall. Changing patterns of temperature and rainfall are leading to diverse responses of migratory and nomadic birds.

## Australia and New Zealand

In the Southern Hemisphere, this is the only region for which relatively good information on the responses of migratory birds is available.

125

## Migration Patterns

All migratory land birds that breed in mainland Australia and Tasmania are intracontinental migrants or migrants from Australia to New Guinea and the islands of Wallacea to the west, all thus being short-distance or partial migrants. Nine species that breed in tropical or subtropical northern Australia are fully migratory to New Guinea and Wallacea during winter. About thirty-nine nonpasserines and fourteen passerines breed in Australia and are partial migrants to New Guinea and Wallacea. Partial migration, including altitudinal shifts, characterizes many additional species of land birds. An estimated fifty species of birds breeding at higher elevations in the Great Dividing Range of eastern Australia show partial shifts to lower elevations during the winter. Some thirty-six species of land birds in interior Australia show variable or irregular movements that depend on year-to-year differences in weather conditions that influence food availability.

Three land bird species, the Shining Bronze-Cuckoo, Long-tailed Koel, and Double-banded Plover, that breed in New Zealand also migrate to island wintering areas—in the western Pacific, in the case of the two cuckoos, or to eastern Australia, in the case of the plover.

## Changes in Migration Patterns

Recent studies show that changes in migration patterns of Australian birds, both of arrival and departure from breeding ranges and of arrival and departure from wintering ranges, are similar in magnitude to those seen in North America and Europe. Most data on changes in migration patterns of Australian birds come from studies in the eastern and extreme western parts of the country. Changes in migration patterns of some Asian nonbreeding visitors have also occurred.

In southeastern Australia, annual maximum temperatures have increased, on average, about $0.17°C$ per decade, and minimum temperatures about $0.13°C$ per decade, since 1960. Annual rainfall has also tended to decline. Dates of arrival and departure of many migratory land birds to and from their breeding ranges in southeastern Australia have changed over the past 30 or so years, according to forty-five data sets covering varying time spans. As in studies of changes in migration timing in the Northern Hemisphere, responses vary among species and localities. Of eighteen species of small land birds for which data were adequate for analysis, ten show significant advances in earliest arrival dates at breeding areas. Nine species, including three cuckoos, a kingfisher, a bee-eater, a pardalote, a whistler, and two monarch flycatchers, showed advances in arrival dates that averaged slightly over a week per decade, based on observation periods ranging from 14 to 34 years in duration (Table 9.1). There was a tendency for the advance in arrival dates to be greatest at the most southern breeding areas. For departure dates from breeding areas in southeastern Australia, four of the eighteen species showed significantly delays, ranging from 10.5 to 21.5 days per decade (Table 9.1). Thus, the four species that advanced arrivals

Table 9.1. Statistically significant changes in spring arrival and fall departure dates of land birds of southeastern Australia in recent decades.

| | Years of Record | Change in Days/Decade[1] | Location |
|---|---|---|---|
| Arrival at Breeding Localities | | | |
| Pallid Cuckoo | 26 | −4.4 | Canberra |
| Common Koel | 34 | −9.1 | Sydney |
| Channel-billed Cuckoo | 29 | −7.7 | Sydney |
| | 17 | −3.7 | N New South Wales |
| | 16 | −12.7 | S New South Wales |
| Sacred Kingfisher | 29 | 4.1 | Canberra |
| | 15 | −15.5 | Sydney |
| Rainbow Bee-eater | 21 | −5.8 | Canberra |
| | 14 | −8.4 | W New South Wales |
| White-faced Gerogyne | 14 | −8.8 | Sydney |
| Rufous Whistler | 19 | −12.2 | Sydney |
| Black-faced Monarch | 26 | −5.3 | Sydney |
| Leaden Flycatcher | 25 | −0.9 | Sydney |
| Departure from Breeding Localities | | | |
| Rufous Whistler | 18 | 19.0 | Sydney |
| Black-faced Monarch | 16 | 21.5 | S New South Wales |
| Leaden Flycatcher | 14 | 15.4 | S New South Wales |
| Rufous Fantail | 14 | 10.5 | S New South Wales |
| Arrival at Wintering Localities | | | |
| White-throated Needletail | 16 | −15.4 | Canberra |
| Departure from Wintering Localities | | | |
| White-throated Needletail | 26 | −5.1 | Sydney |

Source: Data from Beaumont et al. 2006.
[1]Negative values = earlier; positive values = later

and delayed departures lengthened their stay in the breeding range by just under 2 weeks per decade. These tendencies toward earlier arrival and later departure by many species correlated generally with climatic trends toward higher annual maximum and minimum temperatures at their respective localities. One wintering species, a swift that breeds in Siberia and Japan, showed a significantly advanced arrival, averaging over 2 weeks per decade, and earlier departure, averaging just over 5 days per decade.

Climate change is also substantial in Western Australia. Temperatures there have warmed by about 0.8°C since the early 1900s. Since 1950, however, the climate has warmed 0.14°C per decade. At the Eyre Bird Observatory, located on the southern edge of the continent near the Great Australian Bight, maximum temperatures increased significantly between 1984 and 2003. Most of this increase was due to warmer temperatures during the winter months. In far southwestern Australia, the number of hot days increased and cold nights decreased during spring between 1973 and 2000.

Rainfall in southwestern Western Australia has also decreased by 15–20 percent since the mid-1970s, largely in the late fall and winter months. Near Perth, on the southwest coast, runoff into reservoirs decreased by about two-thirds over the period 1911 to 2004. In addition, in the far southwestern corner of Australia, the number of rainy days declined significantly between 1973 and 2000.

At the Eyre Bird Observatory two nonbreeding land birds, the Purple-crowned Lorikeet and the Grey Fantail flycatcher, showed significantly earlier arrival over the period 1984–2003. Lorikeets especially tended to arrive early in years with warmer nighttime temperatures. Three other species also showed tendencies to arrive earlier over the years. The biological basis of these changes was not entirely clear, although the movements of lorikeets may be influenced by flowering patterns of eucalypts. The Fork-tailed Swift, a nonbreeding visitor that breeds in eastern Asia, arrived earlier and departed earlier in warmer years.

Analysis of data on arrival and departure dates for nineteen migratory bird species at the Middlesex Field Study Centre near Manjimup, in southwestern Australia, between 1973 and 2000 showed significantly altered schedules for fourteen species. Ten species, including both land and waterbirds, showed significant trends toward earlier or later arrival, and seven toward earlier or later departure. Among land birds the Slender-billed and White-tailed Black Cockatoos, Red-capped Parrot, and Shining Bronze-Cuckoo showed trends toward earlier arrivals. On the other hand, the Restless Flycatcher and Little Grassbird showed trends toward later arrival. The Red-capped Parrot and Pallid Cuckoo showed trends toward earlier departure. The Shining Bronze-Cuckoo tended to extend its summer stay, but the Striated Pardalote and Little Grassbird significantly shortened their summer presence. Species that tended to arrive in spring generally tended to arrive earlier, while those arriving in fall and winter tended to arrive later. In general, however, the changes in migration schedules were weaker than those noted in southeastern Australia.

Arrival and departure dates and lengths of stay of land birds at the Middlesex Centre showed varied correlations with temperature and rainfall. High minimum temperatures, for example, were correlated with shorter stays by Pallid Cuckoos and Little Grassbirds. The number of rainy days favored later arrival of the Shining Bronze-Cuckoo and also shorter spring and summer stays. Thus, species were highly individualistic in their responses to temperature and rainfall conditions.

In eastern Australia, altitudinal migration of a number of bird species in the Snowy Mountains of Victoria has changed as snow cover has declined since the mid-1900s. Overall, snow cover has decreased by about 27 percent since the 1960s. Of eleven species of migratory birds breeding above 1500 meters in elevation, all but two showed a tendency for earlier spring arrival at breeding elevations. These included three species of honeyeaters that require flowering shrubs for their feeding: Crescent and Yellow-faced Honeyeaters and the Red Wattlebird. Four species, Flame Robin, Olive Whistler, Richard's Pipit, and Striated Pardalote, are passerines that feed on insects. One raptor, the Australian Kestrel, which requires snow-free areas for hunting, and the Fan-tailed Cuckoo, a nest parasite on other birds, also showed a substantially earlier arrival. The two species for which no tendency for earlier arrival was noted were insectivores, one being a long-distance migrant for which the migration schedule may be related to conditions in distant locations.

Although none of these altitudinal migrants is restricted to habitats above tree line, the projected change in snow conditions, and presumably vegetation, during this century indicates that more extensive change in altitudinal migration patterns is likely. The most severe climate change scenario projects the virtual disappearance of the alpine habitat in the Snowy Mountains. The wintering patterns of some montane birds have also changed. In southeast Queensland, the Noisy Pitta used to winter in the lowlands, but now it is a permanent resident in the mountains, where minimum temperatures have warmed over the last half century.

In New Zealand, the evidence for changes in patterns of bird migration is slimmer. The Welcome Swallow, a species that first appeared in the country in the 1920s, has become widely established in lowland areas and continues to spread into highland areas. Between 1962 and 1995, nesting has tended to occur earlier in the summer, with laying of first eggs advancing about 30 days. This change has occurred during the period of greatest climatic warming in New Zealand.

## South America

The South American migration system is the richest of the Southern Hemisphere systems in number of species, with about 266 latitudinal, intratropical, and altitudinal migrants. Unfortunately, little information is so far available on long-term changes in abundance and migration patterns that can be related to global climate change.

Argentina has experienced warming equal to about 1.0°C over the past century, so changes in ranges of migratory birds are expected. In Buenos Aires, the spring arrival of the Gray-breasted Martin advanced about 37.5 days over the period 1976–2000. The arrival date also showed a strong correlation with the increase in mean maximum temperatures during the spring (August through October). Whether or not the establishment of Argentinian breeding populations of the Barn Swallow, beginning in 1980, and the Cliff Swallow, beginning in 1994, is associated with climatic change is uncertain.

In São Paulo State, Brazil, a few long-term studies show that changes in the presence and abundance of some migratory species have occurred, although the role of climate change is difficult to separate from effects of changing land use. Between 1982 and 2003, some migratory birds declined or disappeared in native Itirapina savannas, which have recently experienced more frequent dry years, possibly due to climatic change. These species include the Sooty Tyrannulet and a number of migratory *Sporophila* seedeaters. Studies of birds in a semi-deciduous woodland from 1982 through 2000 showed that some wintering birds, including the Swallow-tailed Cotinga, Fawn-breasted Tanager, and Glaucous-blue Grosbeak, which were present in wet years of the early 1980s, were largely absent in later years. A number of resident birds also declined or disappeared in the 1990s and 2000s. In addition, three migratory summer visitors, the Purple-winged Ground Dove, Dark-billed Cuckoo, and Black-crowned Tityra, were present in the early 1980s but absent in later years. In recent years, the migratory Picazuro Pigeon has invaded open areas in São Paulo State from drier regions to the north and west.

## Southern Africa

Concern about the relation of climate change to bird populations in southern Africa has been slow to develop. As we noted earlier (Chapter 2), a doubling of atmospheric $CO_2$ will likely lead to warming of 2.5–3.0°C in South Africa, especially in the northern interior region. Although African birds in general are predicted to be particularly vulnerable to climatic change, few observations of changes in the ranges of migratory birds are documented. Large areas of arid and semiarid lands bordering the African tropics are expected to experience higher temperatures, reduced precipitation, increased incidence of severe weather events, and increased wildfire frequency. The El Niño Southern Oscillation system influences southern Africa and is expected to strengthen, as well, leading to more intense wet and dry cycles. Drought conditions prevailed in much of western interior South Africa in the 1980s and early 1990s and returned in the mid-2000s.

Southern Africa, below about 15° S, is a winter home to more species of migratory birds that breed in the Northern Hemisphere than in subequatorial South America or Australasia. About forty-two species of land and freshwater birds from Eurasia winter in southern Africa. Only twenty-eight Northern Hemisphere breeders winter in subequatorial South America and ten species in Australasia. Climate change is thus a factor of concern for these species.

Comparison of range maps in handbooks from the 1970s with maps prepared in the early 2000s shows southward range expansions for fifty-two species of land birds of southern Africa. Fully migratory (sixteen), partially migratory (three), and nomadic (ten) species accounted for 55.8 percent of the species showing these expansions, although these groups represented only 39.4 percent of all birds considered. Most of these species are also habitat generalists. Of the sixteen fully migratory species, fifteen were insectivores or invertebrate feeders. Of these, three are wintering European species: the Common Swift, Icterine Warbler, and Marsh Warbler. Seven are aerial insectivores that breed in South Africa, including the African Palm-Swift and six species of widely distributed swallows. Some partial migrants, including the Crested Barbet, Swallow-tailed Bee-eater, and Black-throated Canary, have also moved southward. Increasing drought conditions in interior South Africa have probably favored southward spread of both wintering European species and intra-African migrants. Some of these expansions have probably been favored by climatic change, but a number of colonists of the Cape Peninsula apparently reflect creation of favorable habitat conditions as a result of human activity.

About thirteen species of nonraptorial land birds of the Karoo region of South Africa are strongly nomadic, with more than twenty others showing localized patterns of nomadism. Strongly nomadic species appear to be especially resilient in the face of climate-driven habitat change because of their strategy of movement over long distances in search of favorable breeding areas. Climatic modeling studies suggest that nomadic species such as some bustards, sandgrouse, sparrow-larks, and the Lark-like Bunting, are likely to see shifts in the regions in which conditions favorable to breeding frequently occur. Small nomadic birds such as the sparrow-larks, in particular, are probably less at risk from climate change in southern Africa than are many resident and migratory species. Their ability to move around and find conditions favorable in a given year, however, may enable them to adjust easily. It appears likely that climate change will lead to substantial shifts in areas utilized by many more nomadic and partially nomadic species.

Climatic modeling suggests that other short-distance migrants of southern Africa are likely to be affected by climatic change. The Blue Swallow, which breeds in the highlands of eastern South Africa, will likely lose some of its breeding range, which may also be shifted farther east. Other endemic migratory birds that breed on mountain slopes or at high elevations are also at risk of habitat loss from climate change. The Yellow-breasted Pipit and Mountain Pipit are judged to be at high risk. The Cape Siskin and Drakensberg Siskin are considered to be at medium to medium-high risk.

## Summary

In Australia, where significant climatic warming has occurred, many migratory species in southeastern and southwestern regions show advances in arrival dates and delays in departure dates. In the Snowy Mountains, altitudinal migrants have shown earlier spring arrivals. In New Zealand, the Welcome Swallow has advanced its nesting by about a month. Although many bird species of southern South America are migratory, changes in breeding patterns of only a few species of swallows are documented. Some migratory species may have disappeared from southern Brazil because of the spread of drier conditions. In southern Africa, a number of migratory, nomadic, and wintering species have expanded their ranges southward since the 1970s.

## KEY REFERENCES

Beaumont, L. J., I. A. W. McAllen, and L. Hughes. 2006. "A matter of timing: Changes in the first date of arrival and last date of departure of Australian migratory birds." *Global Change Biology* 12:1339–1354.

Chambers, L. E. 2008. "Trends in timing of migration of south-western Australian birds and their relationship to climate." *Emu* 108:1–14.

Dingle, H. 2004. "The Australo–Papuan bird migration system: Another consequence of Wallace's Line." *Emu* 104:95–108.

Dingle, H. 2008. "Bird migration in the southern hemisphere: A review comparing continents." *Emu* 108:341–359.

Erasmus, B. F. N., A. S. Van Jaarsveld, S. L. Chown, M. Kshatriya, and K. J. Wessels. 2002. "Vulnerability of South African animal taxa to climate change." *Global Change Biology* 8:679–693.

Green, K. and C. M. Pickering. 2002. A scenario for mammal and bird diversity in the Australian Snowy Mountains in relation to climate change. Pp. 241–249 in C. Körner and E. M. Spehn (Eds.), *Mountain Biodiversity: A Global Assessment*. Parthenon Publishing, London.

Hockey, P. A. R., H. A. B. Salata, and A. R. Ridley. 2010. "Non-random responses of African birds to climate change." *Biology Letters* in press.

Jahn, A. E., D. J. Levey, J. E. Johnson, A. M. Mamani, and S. E. Davis. 2006. "Towards a mechanistic interpretation of bird migration in South America." *Hornero* 21:99–108.

Simmons, R., P. Barnard, W. R. J. Dean, G. F. Midgley, W. Thuiller, and G. Hughes. 2005. "Climate change and birds: Perspectives and prospects from southern Africa." *Ostrich* 75:95–308.

Willis, E. O. 2004. "Birds of a habitat spectrum in the Itirapina savanna, São Paulo, Brazil." *Brazilian Journal of Biology* 64:901–910.

# Chapter 10

## *Tropical Land Birds*

Many tropical bird species exhibit seasonal migrations. The frequency and extent of these movements are only beginning to be understood, but they likely involve many species. Most tropical migrants are species that show seasonal altitudinal shifts or short-distance movements at lower elevations that track seasonal patterns of rainfall and food resource availability. In Africa, especially, many species tend to follow the seasonal shift in wet seasons between the Tropics of Cancer and Capricorn. Altitudinal migration is especially frequent, although still poorly documented, in tropical mountains.

Although some species may rely on changing photoperiod to calibrate their annual physiology, for many, rainfall or food conditions are likely to be the direct cues to which the brain physiology of tropical birds is attuned. Many tropical migrants feed heavily on fruit or nectar, which vary seasonally and spatially in almost all tropical habitats. Since these aspects of tropical environments are being affected by global climate change, altitudinal and intratropical migrants are certain to be influenced by the climatic and environmental changes noted in Chapters 2 and 3.

## Tropical Migrants: Mexico, Central America, and South America

Altitudinal migration is common in mountain regions throughout the New World tropics. In southern Durango, Mexico, for example, fourteen species of birds breeding in the oak–pine woodlands or moist ravines of the Sierra Madre Occidental migrate to the Pacific lowlands in winter. These include several species recognized as latitudinal migrants farther north in western North America, plus several species that do not show strong patterns of latitudinal migration, such as the Elegant Trogon, Eared Trogon, Tufted Flycatcher, and Slate-throated Redstart. Farther south, about a quarter of the breeding birds of a cloud forest area in Oaxaca, Mexico, carry out some altitudinal movements. The Black-headed Nightingale-Thrush and Ochre-bellied Flycatcher show altitudinal migration in Nicaragua.

Altitudinal movements are well developed and are documented in greater detail in Costa Rica than anywhere else. Many species of Costa Rican birds migrate between seasonal forests of the Atlantic and Pacific lowlands and high montane forests. About 30 percent of the breeding birds of forests on the Atlantic slope show altitudinal migration, most breeding at higher elevations than those occupied during the nonbreeding season. Between 50 and 75 species, especially frugivores and nectarivores, show at least partial altitudinal migration, with movements into higher elevations being primarily in the dry season. A detailed analysis for the Braulio Carrillo National Park area of Costa Rica, which spans an elevational gradient from 50 to over 2000 meters, characterized 69 species of tropical birds as showing altitudinal migration. The largest numbers were frugivores (32 species), nectarivores (18 species), and feeders on small insects (14 species). A number of tyrannid flycatchers (7 species) also were characterized as being intratropical migrants. Another survey of forest birds along an altitudinal transect through Braulio Carrillo Park to the La Selva Biological Station at the base of the mountains on the Atlantic slope showed that 56 of 261 species (21.4 percent) observed or captured in mist nets were altitudinal migrants. The relative abundance of altitudinal migrants was greatest at the higher elevations. Most of these species were frugivores. Still another survey, in the Tilarán Mountains farther west, showed that 17 percent of all species encountered were altitudinal migrants. These migrants included species breeding at all levels in the mountains.

Altitudinal migrations of several Costa Rican cloud forest birds are known in detail. Resplendent Quetzals, for example, spend nonbreeding periods in seasonal moist forests of the Pacific slope, where they take advantage of fruiting of tree species of these moist lowlands. During the breeding season, they migrate upward to cloud forests. The quetzal is an obligate frugivore and feeds on at least forty-one plant species in the Monteverde Cloud Forest Reserve, Costa Rica. It depends largely on fruits of about eighteen species of woody plants of the Lauraceae, and the distributions and fruiting schedules of these plants control the seasonal altitudinal movements of these birds. Because of widespread deforestation, especially in its nonbreeding habitats, quetzal populations are declining almost everywhere. In Costa Rica, the main problem is the fragmentation and destruction of forests used by the birds in the nonbreeding season.

The Three-wattled Bellbird has one of the most remarkable migration patterns yet documented for a tropical bird. In Costa Rica, one population breeds during March through May at elevations of 1000–1800 meters on the Atlantic slope of the Tilarán Mountains. After breeding, they cross onto the Pacific slope of these mountains, where they remain for about 3 months. In September and October they migrate to the lowland forests of southeastern Nicaragua and northeastern Costa Rica. In November and December, they return across the mountains to forests of the Pacific coast of Costa Rica, staying there until March, when they begin their return to breeding areas.

The Bare-necked Umbrellabird is another large frugivore that breeds in upper elevations on the Tilarán Mountains in Costa Rica. These birds nest at elevations of about 1400 meters during the months of April through July, which coincides with the period of maximum abundance of fruits utilized by the species at this elevation. Following breeding, the birds move to lower elevations on the Atlantic slope, and from September through November, they occur mainly in premontane rain forest at an elevation of about 400 meters. At this season, the production of fruits similar to those utilized in the high mountains is high at these lower elevations. Umbrellabirds probably spent portions of the nonbreeding season in lowland forests at even lower elevations. Extensive clearing of these forests has reduced the area of suitable nonbreeding habitat substantially and has probably contributed to the decline in numbers of this endemic Central American species.

Black Guans, Crested Guans, Emerald Toucanets, and several other large frugivores show similar patterns of altitudinal migration in the Tilarán Mountains. Nonbreeding movements of most of these species, except for the Crested Guan, do not reach elevations as low as those of the umbrellabird, however. In other parts of Central America, extensive seasonal movements are known for the Mealy Parrot and at least two species of macaws.

A number of smaller frugivorous birds also show altitudinal movements. In Costa Rica, these include the White-ruffed Manakin, the White-crowned Manakin, and possibly the Red-capped Manakin. The Red-capped Manakin also appears to show seasonal habitat shifts. In Veracruz,

Mexico, the Common Bush Tanager shows periodic altitudinal movements. Several workers have also documented localized altitudinal movements of tropical birds in mountain areas in response to severe weather or food resource failures.

Many tropical nectar-feeding birds also are altitudinal migrants. At least 24 species of hummingbirds show altitudinal migration in Costa Rica, correlated with flowering of their preferred plants. These include the Purple-throated Mountain-gem, Green-crowned Brilliant, Green Violetear, Brown Violetear, White-crested Coquette, Black-crested Coquette, Green Thorntail, and Snowcap.

Altitudinal and intratropical migration patterns are still inadequately documented in South America. In western Colombia, however, about 17 percent of the tropical breeding birds, including twelve of twenty-two species of hummingbirds, are short-distance migrants. Other migrants include species of parrots, pigeons, quetzals, cotingas, flycatchers, thrushes, and tanagers. Substantial altitudinal migration occurs by cloud forest birds, especially hummingbirds and frugivores, in both western Colombia and southeastern Brazil. Recent studies suggest that altitudinal migration may not be as well developed along the eastern Andean slope in southeastern Peru.

Latitudinal movements are also shown within the tropics by a number of New World species, such as the Yellow-green Vireo and Piratic Flycatcher, which breed from southern Mexico southward through Central America and winter in northern South America. Farther south, intratropical migration is shown by several species of Tyrannid flycatchers.

Migration and other life cycle processes in the Yellow-green Vireo apparently have a strong endogenous control, but even within the tropics, photoperiod changes may serve to time these events. Natural selection may therefore be able to adjust life cycle timing, as well as breeding and non-breeding ranges, as climate change occurs.

## Tropical Migrants: Africa

Altitudinal migration is also frequent in Africa. Altitudinal movements of Tanzanian forest birds occur between montane forests of the Eastern Arc Mountains and lower-elevation or coastal forests. At least twenty-six species of birds that breed in highland forests move downward during cold, dry weather from June to September. The species involved are a diverse assem-

TABLE 10.1. Montane birds of the Eastern Arc Mountains of eastern Africa recorded in foothill and coastal forests during the cold, dry season from June to September.

| | Foothill[1] | Coastal[1] |
|---|:---:|:---:|
| White-chested Alethe (*Alethe fuelleborni*) | ** | * |
| White-starred Robin (*Pogonocichla stellata*) | ** | * |
| Orange Ground-Thrush (*Zoothera gurneyi*) | ** | * |
| Evergreen Forest Warbler (*Bradypterus mariae*) | ** | |
| Stripe-cheeked Greenbul (*Andropadus milanjensis*) | * | * |
| Lemon Dove (*Aplopelia larvata*) | * | * |
| Rameron Pigeon (*Columba arquatrix*) | * | |
| Bar-tailed Trogon (*Apaloderma vittatum*) | * | |
| Spot-throat (*Modulatrix stictigula*) | * | |
| Black-fronted Bushshrike (*Telephorus nigrifrons*) | * | |
| Barred Long-tailed Cuckoo (*Cercococcyx montanus*) | * | * |
| White-tailed Crested-Flycatcher (*Elminia albonotata*) | * | |
| Kenrick's Starling (*Poeoptera kenricki*) | * | |
| Bronze-naped Pigeon (*Columba iriditorques*) | * | * |
| Banded Sunbird (*Anthreptes rubritorques*) | * | |
| Shelley's Greenbul (*Andropadus masukuensis*) | * | |
| Western Mountain-Geenbul (*Andropadus tephrolaemus*) | * | |
| Sharpe's Akalat (*Sheppardia sharpei*) | * | |
| Mountain Wagtail (*Motacilla clara*) | * | |
| Fuelleborn's Boubou (*Laniarius fuelleborni*) | * | |
| Waller's Starling (*Onychognathus walleri*) | * | |
| Purple-throated Cuckoo-Shrike (*Campephaga quiscalina*) | * | |
| Gray Cuckoo-Shrike (*Coracina caesia*) | * | * |
| Red-faced Crimsonwing (*Cryptospiza reichnovii*) | * | |
| Olive Woodpecker (*Dendropicos griseocephalus*) | * | |
| African Hill Babbler (*Pseudoalcippe abyssinica*) | * | |

*Source:* Modified from Burgess and Mlingwa 2000.
[1]Key: ** = Most frequent; * = less frequent

blage of tropical forest birds (Table 10.1). All of these species occur seasonally in foothill forests at elevations below 800 meters. Of these, eight species appear seasonally 100–240 kilometers east of the mountains in forests of the coastal lowlands. In addition, all twenty-six species appear to be partial migrants, with some individuals remaining at the higher-elevation breeding areas. The birds return to the mountain forests when the rains begin in November and December, which is the beginning of the breeding period for most species.

In the Udzungwa Mountains of south-central Tanzania, one unit of the Eastern Arc chain, an additional four species appear to decline in abundance at high elevations during the cold season from late February through September, indicating substantial altitudinal movements. These are the Eastern Olive Sunbird, African Paradise-Flycatcher, Klass's Cuckoo, and African Broadbill.

Large frugivores in other parts of Africa also show patterns of long-distance movement, correlated with fruit availability. Seasonal movement between forest areas with different seasonal patterns of fruit availability are shown by Black-casqued and White-thighed Hornbills in Cameroon, west Africa. Radio tracking has shown that many of these birds move up to 150 kilometers within a single month and 290 kilometers over 3 months.

Many other African species show intratropical migrations correlated with wet and dry season patterns. These include several bee-eaters, rollers, nightjars, pipits, starlings, and other species. Nine species of bee-eaters show some pattern of migration. The White-throated Bee-eater breeds along the southern edge of the Sahara and in parts of eastern Africa, wintering in more-southern rain forest and savanna areas. The Southern Carmine Bee-eater exhibits separate breeding populations in central and southern Africa. After breeding, northern birds move south, and southern breeders disperse both north and south of their breeding range. Five species of rollers are partial intratropical migrants. Several nightjars are partial migrants, while the Pennant-winged Nightjar is a transequatorial migrant, breeding in southern Africa and wintering in central Africa.

## Tropical Migrants: Southeast Asia

Patterns of long-distance movement are also shown by hornbills and other large frugivores in the East Indies. In Borneo, these involve large seasonal movements of hornbills of the genus *Aceros* and other frugivores, correlated with changes in fruit availability. Populations of the Wreathed Hornbill also wander widely and even cross ocean gaps such as that between Bali and Java. On Sulawesi, the abundance and flock size of the Knobbed Hornbill varies with abundance of fig fruit, suggesting that this species also tracks seasonal patterns of fruit availability.

## Tropical Migrants: Australia and New Guinea

In New Guinea, the Dwarf Cassowary, although flightless, shows altitudinal movements as it tracks the availability of different fruits. These birds are apparently active over an altitudinal range from 550 to 1450 meters, with most of the migratory individuals being females.

## Tropical Migrants: Oceanic Islands

In Hawaii, the Iiwi and Apapane move altitudinally from wet forest areas to subalpine woodlands and dry woodlands on the leeward side of the island of Hawaii in response to nectar availability patterns. Large altitudinal movements of these species also occur on the windward side of this island.

## Impacts of Global Change on Migratory Birds in the Tropics

As we noted in Chapter 2, mountain areas in both the Northern and Southern hemispheres are experiencing climatic warming. Several modeling studies suggest that tropical cloud forest climates are likely to retreat upslope as a result. One general circulation model predicts that under a doubled $CO_2$ regime, the base of orographic cloud systems that create cloud forest climates in several world regions, including Costa Rica, will be pushed upward by 200 meters or more during the dry season. This effect results from warming of the ocean surface and the pattern of interaction of temperature and humidity as air masses flow onto the mountains. These changes will have a strong drying effect on epiphyte-rich cloud forest canopies in Costa Rica. Along the eastern flank of the tropical Andes in South America, a rise of 600 meters in the lower limit of cloud forest is predicted during this century, which would greatly restrict this critical avian environment.

Other modeling studies suggest that deforestation, especially in lowland areas, is also contributing to climatic change in Costa Rica. Simulations of cloud conditions over the Tilarán Mountains of Costa Rica under conditions of varying lowland deforestation show that when the lowland and foothills are forested, orographic clouds strike the mountains at the lowest elevations. They cover the largest area of the mountains, and they remain for the longest time. For the current level of deforestation, the montane forest area immersed in cloud was reduced by 5–13 percent, and the base of orographic clouds in the afternoon was raised by 25–75 meters. Continued deforestation was predicted to cause a further 15 percent decrease in the area immersed in clouds and to raise the cloud base by an additional 125 meters. The simulations suggest that lowland deforestation heats the air and lowers the dew point of air masses that flow from lowlands into the mountains and give rise to orographic clouds.

In Costa Rica, major climatic changes have occurred in the cloud forest zone of the Tilarán Mountains, where the Monteverde Cloud Forest Reserve is located (see Chapter 3). The drying conditions that now exist are correlated with major biological changes, including the disappearance of twenty species of amphibians and the colonization of higher elevations by at least fifteen species of birds. For example, on a visit to the Monteverde Reserve in 2005, I found that the Golden-crowned Warbler, which first nested there in 1994, is now a common species. Furthermore, several altitudinal migrants that breed at Monteverde, including the Resplendent Quetzal, Fiery-throated Hummingbird, Collared Redstart, and Yellow-thighed Finch, have declined in abundance or retreated to higher elevations.

Climatic change, due to greenhouse warming and human modification of tropical forests and other natural communities, also poses a serious threat to other altitudinal migrants. Warming of tropical lowlands, together with forest clearing, may alter physical and biotic conditions of low-elevation areas used by altitudinal migrants in the non-breeding season. In Costa Rica, for example, Three-wattled Bellbirds are dependent on Pacific-slope mid- and low-elevation forests during the nonbreeding season. Lowland areas have warmed by about 0.25°C per decade since the mid-1970s, and their forests are being rapidly destroyed, so bellbirds are forced to utilize scattered forest remnants and second-growth stands. Healthy lowland areas such as Manuel Antonio National Park on the Pacific coast, where I heard nonbreeding birds calling in 2005, are essential to the survival of populations of this species. Records of calling birds in breeding areas in the Tilarán Mountains suggest that bellbird populations may be only about 20 percent of their numbers in the early 1980s. Similar concerns exist for other altitudinal migrants, such as umbrellabirds, large parrots, macaws, and other large frugivores breeding in the mountains.

Afro-montane areas are also experiencing environmental change. On Mount Kilimanjaro in East Africa, the upper forest line has dropped by several hundred meters over the last century. The warming climate has led to an increase in frequency and intensity of fires at high elevations, causing downward retreat of the forest. In Madagascar, recent climatic warming has led to upward distributional shifts in the many reptiles and amphibians, but responses of birds that show altitudinal movements have yet to be documented.

Analysis of elevational ranges of southeast Asian birds as reported in 1971 and 1999 in field guides suggests that many resident or locally migrant species have moved upward in elevation. At least one altitudinal migrant, the Little Forktail, has shifted both lower and upper elevational limits to higher elevations. Circumstantial evidence suggests that these shifts are largely due to climate change.

As we noted in Chapter 3, even a 1.0°C warming will cause a substantial reduction in range of many rain forest animals in northern Queensland, Australia. The Queensland mountains contain thirteen endemic species, ten of which are confined to cooler montane areas. Among the latter is the Golden Bowerbird, a partial altitudinal migrant. Bowerbirds breed at elevations of 700–1500 meters. Modeling studies show that the bowerbird is highly sensitive to the climatic change that is likely to occur in the early part of the twenty-first century. Climatic warming of 1.0°C will likely reduce the breeding habitat by 50 percent, while warming of 3.0°C will restrict breeding habitat to very small patches of forest at the highest elevations of two mountain ranges. These small breeding subpopulations would be much more isolated from each other and subject to a high risk of extinction. With warming greater than 3.0°C, suitable breeding habitat would not exist.

In Hawaii, the ranges and movements of the Iiwi and Apapane, as well as those of other endemic Hawaiian birds, are likely to be affected by global warming. Temperature zones in the Hawaiian Islands are likely to be displaced upward by 360–450 meters, given a warming of 2–2.5°C at sea level.

## Summary

Altitudinal migration is frequent in fruit- and nectar-feeding birds in mountainous areas from southern Mexico through Central and South America. In many of these areas, the cloud forest zone is also being pushed to higher elevation by climatic warming and drying. Similar patterns of migration and environmental change occur in Hawaii, East Africa, southeast Asia, and northern Australia. Based on detailed observations in Costa Rica and other New World localities, many altitudinal migrants are probably altering their migration patterns. In west Africa and southeast Asia, several species of hornbills, and probably other large frugivores, also show substantial seasonal movements. Although a few impacts of global change on these migratory species are just beginning to be documented, changing climate has likely contributed to population declines of some species.

## KEY REFERENCES

Blake, J. G. and B. A. Loiselle. 2000. "Diversity of birds along an elevational gradient in the Cordillera Central Costa Rica." *The Auk* 117:663–686.

Burgess, N. D. and C. O. F. Mlingwa. 2000. "Evidence for altitudinal migration of forest birds between montane Eastern Arc and lowland forests in East Africa." *Ostrich* 71:184–190.

Chaves-Campos, J. 2004. "Elevational movements of large frugivorous birds and temporal variation in abundance of fruits along an elevational gradient." *Ornitología Neotropical* 15:433–445.

Colwell, R. K., G. Brehm, C. L. Cardelus, A. C. Gilman, and J. T. Longino. 2008. "Global warming, elevational range shifts, and lowland biotic attrition in the wet tropics." *Science* 332:258–261.

Hilbert, D. W., M. Bradford, T. Parker, and D. A. Westcott. 2004. "Golden bowerbird (*Prionodura newtoniana*) habitat in past, present and future climates: Predicted extinction of a vertebrate in tropical highlands due to global warming." *Biological Conservation* 116:367–377.

Holbrook, K. M., T. B. Smith, and B. D. Hardesty. 2002. "Implications of long-distance movements of frugivorous rainforest hornbills." *Ecography* 25:745–749.

Peh, K. S.-H. 2007. "Potential effects of climate change on elevational distributions of tropical birds in southeast Asia." *Condor* 109:437–441.

Pounds, J. A., M. P. L. Fogden, and J. H. Campbell. 1999. "Biological response to climate change on a tropical mountain." *Nature* 398:611–615.

Pounds, J. A., M. P. L. Fogden, and K. L. Masters. 2005. Responses of natural communities to climate change in a highland tropical forest. Pp. 70–74 in T. E. Lovejoy and L. Hanna (Eds.), *Climate Change and Biodiversity*. Yale University Press, New Haven, CT.

Williams, S. E., E. E. Bolitho, and S. Fox. 2003. "Climate change in Australian tropical forests: An impending environmental catastrophe." *Proceedings of the Royal Society of London B* 270:1887–1892.

# Chapter 11

# *Raptors*

Raptorial birds, including members of the orders Falconiformes (hawks, falcons, vultures, and relatives) and Strigiformes (owls) are among the birds most vulnerable to human influences. Their positions at high levels in the food chain place them at risk due to influences as diverse as direct persecution, human modification of habitat conditions and prey abundance, pollution by bioactive chemicals, and climatic change that alters habitat and prey availability. Long-distance migratory species are considered to be at special risk from combinations of these factors.

Not surprisingly, distinguishing the effects of climatic change itself on raptors is difficult. Overall, climatic warming in the temperate zones should favor permanent resident species and partial migrants that can transition to permanent residency. As a result, long-distance migrant raptors may be faced with increased competition on their breeding ranges. Trends in population dynamics are evident for many species, and climate change is likely to be at least a contributing factor for some of these.

## North America

In North America, the ranges of some species have shifted, the populations of some species have increased and others decreased, and migration schedules and breeding seasons of a few species have changed.

## Ranges

In eastern North America, several raptors appear to be spreading their breeding and wintering ranges northward, perhaps in response to climatic warming. The Turkey Vulture, which uses both sight and smell to locate dead animals, may have benefited from warmer weather and reduced snow cover. It spread northward in the upper Midwest and Canadian Prairie Provinces in the latter part of the 1900s. The Black Vulture, which apparently does not use a sense of smell to locate food, has expanded its breeding range northward somewhat less since the 1930s. It now breeds north to Pennsylvania and New Jersey.

Other raptors spreading northward in North America may include the Mississippi Kite, which nested in southern Ohio in 2007. An old record of the species from Ohio came from remains in an archeological site, perhaps not from a bird that lived in the state. Since the 1970s, this species has also expanded its range in Oklahoma, Kansas, Texas, New Mexico, and Colorado. The Broad-winged Hawk also appears to be spreading westward in Canada, breeding as far west as northeastern British Columbia.

Likewise, the winter range of some species seems to be moving northward. Between 1993 and 2006, for example, numbers of wintering Rough-legged Hawks were lower in eastern, western, and southern parts of their winter range and significantly greater in the northern plains region, compared with 1979–1992. This shift correlated with reduced snow cover in the areas with fewer Rough-legged Hawks, and also with a northward shift of Red-tailed Hawks, a potential competitor. These shifts appear to reflect climatic warming and milder winter conditions. Smaller numbers of Northern Harriers and Sharp-shinned Hawks are being recorded at raptor observation sites around the Gulf of Mexico, and numbers are larger in winter counts farther north. These species also seem to be wintering at higher latitudes.

Projections of future climate change suggest that many more range changes are likely, based on modeling of the breeding ranges and abundance patterns of a number of raptors in the eastern United States under a scenario of climate change with a doubled $CO_2$ concentration. These projections were based on two models, the Canadian Centre for Climate model and the Hadley Center for Climate Prediction and Research model. Data from the North American Breeding Bird Survey (BBS) were used to model the responses of individual species. Eight raptors were considered. Both the Turkey Vulture and the Black Vulture were predicted to increase in abundance in the more southern portions of their present ranges. The Turkey Vulture was predicted to expand northward to a limited extent and the Black Vulture to expand northward substantially. The Mississippi Kite, now mostly confined to the southern tier of states, was predicted to spread northward over much of the northeastern United States and disappear from much of its southern range. The breeding range of the Northern Harrier was predicted to retreat northwestward, largely disappearing from New England and the lower Midwest. More limited changes in range and abundance are predicted for the Red-tailed Hawk, Red-shouldered Hawk, and Broad-winged Hawk. The American Kestrel was also projected to show patterns of increase and decrease in different portions of its current range. Another study, which combined habitat and bird occurrence models with models of future climate change in northeastern North America, also predicted substantial decreases in abundance of Northern Harriers and American Kestrels.

The ranges of some migratory owls may have changed as a result of climatic warming. The Barred Owl, a partial migrant in its most northern areas, expanded its range westward across southern Canada beginning in the early 1900s and contacted the range of the Spotted Owl in the Pacific Northwest in the early 1970s. Substantial summer warming occurred in the forest region of central Canada during this period, suggesting that this warming, whether natural or anthropogenic, has favored Barred Owl reproduction. Landscape changes due to human activities may also have contributed to range expansion by the Barred Owl and perhaps other raptors in this region, such as the Broad-winged Hawk.

## Populations

Population trends for raptors are largely based on the BBS, the Christmas Bird Count (CBC), and counts at observation sites where migrating birds are concentrated. BBS data indicate that since 1966, Black Vultures have increased over 3 percent annually, with the rate of increase equaling 2.5 percent since 1980. Breeding Turkey Vulture numbers in North America have also risen over 2 percent annually since 1980. Later BBS data indicate that since 1980, annual increase has exceeded 12 percent. Christmas Bird Count data suggest that from 1990 to 2003, numbers of Black Vultures increased almost 6 percent annually and numbers of Turkey Vultures about 1.8 percent annually. Although the numbers and distributions of vultures have fluctuated in response to a number of historical human influences, the widespread increase in abundance and range appears most likely to reflect climatic change.

In Canada, BBS and CBC data, counts at HawkWatch migration sites, and other data show that increases in abundance of the Turkey Vulture have occurred continent-wide. Declines are evident, however, in populations of Northern Harriers, populations and reproductive success of Swainson's Hawks, and populations of Short-eared Owls in the plains regions of western Canada. These did not appear to be fully explained by agricultural change and change in prey population dynamics. The Rough-legged Hawk, an Arctic and subarctic breeder that feeds primarily on small rodents, has also shown a general decline in numbers over the past 30 years in the northern United States and Canada.

For northeastern North America, trends in numbers of raptors counted on migration from 1974 through 2004 are available for three subregions: Atlantic coast, Great Lakes, and Inland areas. Species showing population increases throughout the region included the Osprey, Bald Eagle, Cooper's Hawk, Merlin, and Peregrine Falcon. The Golden Eagle showed increases in Great Lakes and Inland areas. Long-term declines were seen for the Broad-winged Hawk and Red-tailed Hawk in parts of the northeastern region.

The American Kestrel has shown population declines at many migration monitoring sites, on BBS routes, and at CBC sites in North America since 1974. The greatest decline has been in the northeastern United States and eastern Canada, with the western United States also showing a serious decline. These declines have accelerated at most migration monitoring sites since 1994. The causes of these declines are uncertain, but climatic change cannot be ruled out as a contributor.

Analysis of BBS and CBC data and counts at HawkWatch sites in western North America show a number of trends in populations of raptors from the mid-1980s to early 2000s. Some species appear to have increased. The Turkey Vulture showed a general increase and northward range expansion through about 1998, paralleling the pattern seen in eastern North America. Climatic warming, as well as changes in human activities, may have contributed to this pattern. Many other species appear to have declined during drought conditions in the 1990s, with some recovery since the late 1990s. Northern Harrier populations appeared to increase in the 1980s and 1990s but have declined since the late 1990s. The Broad-winged Hawk, which is expanding its breeding range in western Canada, appears to be increasing in most areas. In the western United States, counts at migration monitoring sites and CBC sites also show declines in American Kestrels since about 1995. In all likelihood, more than one factor has contributed to American Kestrel declines.

Over the period 1993–2004, observations of migrating raptors in Veracruz, Mexico, in spring and fall provide data for five species that give estimates of population trends. The Turkey Vulture showed an annual increase of 5.9 percent per year over this period. Swainson's Hawk showed an annual increase of 11.1 percent, Broad-winged Hawk 2.9 percent, Osprey 3.2 percent, and Cooper's Hawk 1.6 percent.

Many of the trends of increasing populations likely reflect recovery of populations from the pesticide-related declines of the late 1960s and 1970s. Other trends may reflect the influence of drought episodes on prey populations, especially in western North America, and trends in vegetation structure related to human activities.

For some species, shifts in wintering distribution may reflect climatic changes. During the 1980s and early 1990s, migrating Sharp-shinned Hawks were recorded in declining numbers at many fall observation sites in eastern North America. The apparent decline was greatest at sites close to the Atlantic coast. Analysis of CBC data over this general time period indicated that the numbers of wintering hawks in New England increased considerably. The numbers of hawks recorded at sites such as Cape May, New Jersey, have declined as the numbers recorded in CBCs in New England have increased. In Florida, in addition, the numbers recorded declined appreciably. These observations suggest that Sharp-shinned Hawks were wintering farther north during a period in which much warmer winter weather has predominated. Whether this shift in distribution of the early winter hawk population is directly related to climatic factors or to their small-bird prey abundances is uncertain.

Few data are available on owl population changes in North America. BBS data suggest long-term declines for the Short-eared Owl in Canada and the United States. The northern subspecies of the Ferruginous Pygmy Owl, perhaps a partial migrant, is experiencing a serious population decline in Sonora, Mexico. In southern Arizona, the species has also decreased in abundance since the mid-1900s. Dry winters related to global climate change may be a contributor. The Burrowing Owl, a partial migrant, has disappeared from significant portions of its original range in the central Great Plains, Canadian Prairie Provinces, and Pacific Northwest. Based on BBS and CBC data, populations also appear to be declining in many states and provinces, although the apparent causes appear to be related to habitat conditions.

## *Migration Patterns*

Migration patterns have also changed for some short-distance migrant raptors. In Worcester County, Massachusetts, for example, the Turkey Vulture advanced its spring arrival date by about 55 days during the period 1932–1993. Between 1951 and 1993, in the Cayuga Lake Basin in New York State, this species on average arrived almost a month earlier in spring than in the period 1903 to 1950. On the other hand, there appears to be essentially no difference in spring arrival dates of Broad-winged Hawks, a long-distance migrant, in northern Maine over the past century.

## *Breeding Seasons*

In the western United States, some migratory owls have begun to nest earlier in the season. In Arizona, the Elf Owl and Flammulated Owl have advanced their breeding dates, correlated with climatic warming, as have other small resident owls. In Colorado, Flammulated Owls advanced their breeding by almost 8 days between 1981 and 2006. This advance is closely correlated with increases in spring temperatures in their breeding habitat.

# Eurasia

In northern Europe, a number of raptors have shown range and population changes likely related to climatic warming. In Finland, both northward and southward range shifts by a number of raptors occurred over a 12-year period in the late 1900s. Among hawks and falcons with distributions in more southerly parts of the country, three species, the Eurasian Hobby, Western Marsh Harrier, and Eurasian Buzzard, showed significant range expansions northward. Two others, the Goshawk and Honey Buzzard, tended to retreat southward. The Rough-legged Hawk, a species with a more northern range, also tended to retreat southward. Among owls, most species showed southward shifts in their northern limits, except for the Boreal Owl, which expanded its range northward.

The declines of populations of some northern hawks appear to be due to reduced abundance of their prey. In northern Scandinavia, populations of the Northern Harrier and Rough-legged Hawk have declined, based on numbers of migrating birds at Falsterbo, southern Sweden. These seem to reflect lower populations of their tundra prey that have become less cyclical and somewhat stabilized at lower densities because of warmer winters resulting from the prolonged influence of the positive phase of the North Atlantic Oscillation (NAO). For the Rough-legged Hawk, the decline in numbers may also reflect in part the range change noted in Finland. Climatic warming appears likely to reduce the northern Scandinavian habitat of this species by 85–96 percent by 2051–2080.

Six species of raptors that migrate from the western Palearctic to sub-Saharan Africa showed significant declines in populations between 1970 and 1990 and/or between 1990 and 2000. These species were the Black Kite, Egyptian Vulture, Pallid Harrier, Lesser Kestrel, Levant Sparrowhawk, and Red-footed Falcon. In contrast, only three species showed significant tendencies to increase in abundance: Montagu's Harrier, Booted Eagle, and Osprey. Five additional species showed no significant trends in either period. Thus, the responses of birds in this group are highly individualistic, presumably reflecting specific changes in the habitats and prey utilized in the various phases of their migratory cycles.

Some species may also be altering the timing of their spring and fall migrations. In Oxfordshire, England, the Eurasian Hobby has not changed its arrival date in spring, but it delayed its fall departure by over 2 weeks over the period 1971–2000. On the Hanko Peninsula in southernmost Finland, the Rough-legged Hawk showed a significant tendency to arrive earlier in spring during the positive phase of the North Atlantic Oscillation.

Arrival at and departure from winter ranges has also changed for some species. Wintering Merlins and Northern Harriers now arrive significantly earlier on their winter ranges in southeastern England. Merlins also remained on the winter range significantly later, and both species significantly extended their winter range stays.

Breeding seasons are shifting, as well. Near Lake Lugano in northern Italy, the Black Kite advanced its egg-laying date by 10–11 days over the period 1992–2002. The laying date has become earlier as spring temperatures have warmed. In this area, the kite feeds primarily on fish, and the tendency for earlier laying was strongest for pairs nesting close to the lakeshore. The advance in laying date appeared to reflect an advance in the time of peak food availability, suggesting that Black Kites are able to respond quickly to this aspect of climatic change. Although breeding became progressively earlier and clutch size also tended to increase through the years, the number of young fledged per nest has remained essentially constant. Nevertheless, the response of this species shows one of the most rapid adjustments by birds to changing climate. Western Marsh Harriers in the Netherlands also nest earlier and have larger clutches in warmer years.

Weather conditions strongly influence breeding of the Lesser Kestrel, a colonial raptor breeding in southwestern Spain. This widespread species has declined in numbers throughout the western Palearctic and disappeared from parts of its former range in Europe. In Spain, it has declined precipitously since the 1970s. From 1966 through 2000, rainfall in southwestern Spain decreased significantly between February and June, the period of falcon arrival, courtship, and nesting. Rainfall is generally beneficial to reproductive activities, although wet weather can also be detrimental. Demographic modeling of the breeding activities of six colonies from 1988 through 2000, however, did not reveal a pattern of reduced reproductive success related to climatic change.

Northern owls in the Old World may already be experiencing changes in population dynamics. Northern Scandinavian populations of the Boreal Owl, a somewhat nomadic or migratory bird that concentrates its predation on voles, have apparently changed from a 3–4-year cyclical pattern to a pattern of annual fluctuation. The average abundance of owls has also declined. These changes reflect the increasingly mild winters occurring under the influence of the positive phase of the NAO, which result in a shorter period of winter snow cover. In Finland, on the other hand, the range of the Boreal Owl has expanded northward. Decline of Snowy Owl populations in Fennoscandia is also linked to lower populations of lemmings. Some populations of strictly nonmigratory owls, such as the more southern Tawny Owl, on the other hand, seem to show significant positive relationships to the warm NAO phase.

Modeling studies suggest that some other owls of northern Scandinavia will likely lose much of their habitat. In Finland and adjacent Norway, Sweden, and Russia, the Great Gray Owl was projected to lose over 83 percent of breeding habitat by AD 2051–2080, and the Northern Hawk Owl over 94 percent, given global warming of 3.8°C. Under a milder scenario of warming by 2.0°C, habitat loss was only slightly smaller.

## Southern Hemisphere

In Australia, where three-quarters of the twenty-four species of Falconiformes are partial migrants, substantial climatic warming has occurred. Several raptors have shown changes in ranges or migration schedules. The Pacific Baza, for example, is reported to be extending its range southward in eastern Australia. The Australian Kestrel has shown a trend toward earlier arrival in alpine regions of the Snowy Mountains, correlated with earlier snowmelt. Continued climatic change is likely to alter the distributions of other raptors. The Greater Sooty Owl in Victoria is likely to lose a third of its range, given a warming of 3°C.

In South Africa, modeling studies suggest that several groups of raptors will be affected by climatic warming. The Cape Griffon and Lammergeier, both mountain species, are projected to abandon north-facing nesting sites and disappear from northern areas. The Cape Griffon has already disappeared from sites in Namibia and Zimbabwe. Increased thorn woodland density, possibly due to climate change, may interfere with the ability of Cape Griffons to locate carcasses to scavenge. The Jackal Buzzard and Black Harrier are also likely to be forced into smaller, more southern ranges. In South Africa, Black Harriers are also highly sensitive to fragmentation of natural habitat due to expansion of agricultural land.

The Black Goshawk in South Africa has expanded its range southward, colonizing the Cape of Good Hope peninsula over the past century. Although the species utilizes pine plantations for nesting, these sites were available much earlier. Climatic warming, especially during winter, and a longer winter rainy season may contribute to the success of the species.

Drying of the Okavango Swamp in Botswana due to climatic warming and reduced precipitation will likely reduce populations of species such as the African Marsh Harrier and Banded Snake Eagle. Species such as the African Grass Owl, Secretary-bird, and various kites are likely to decline as grasslands are replaced by shrubland and woodland.

## Summary

In North America, Black and Turkey Vultures are increasing in numbers and expanding their breeding and winter ranges northward. Populations of a number of other raptors in Canada and the western United States are also increasing, and some, such as the Sharp-shinned Hawk, may also by wintering farther north. Climate modeling studies suggest that other raptors will follow suit. In Europe, some species appear to be spreading northward, others withdrawing southward. Some raptors are also advancing their spring arrivals. The Black Kite appears to have advanced its nesting period in northern Italy. Some Australian raptors are also moving to higher latitudes and altitudes. Habitat loss due to changing climate will likely reduce populations of several South African raptors.

## KEY REFERENCES

Bildstein, K. L. 2006. *Migrating Raptors of the World: Their Ecology & Conservation*. Cornell University Press, Ithaca, NY.

Bildstein, K. L., J. P. Smith, E. R. Inzunza, and R. R. Veit (Eds.). 2008. *State of North America's Birds of Prey*. Series in Ornithology, No. 3, viii + 466 pp.

Hoffman, S. W. and J. P. Smith. 2003. "Population trends of migratory raptors in western North America, 1977–2001." *The Condor* 105:397–419.

Kiff, L. 2000. The current status of North American vultures. Pp. 175–190 in R. D. Chancellor and B.-U. Meyberg (Eds.), *Raptors at Risk*. World Working Group on Birds of Prey and Owls, Berlin and Hancock House Publishers, Surrey, BC, Canada.

Kirk, D. A. and C. Hyslop. 1997. "Population status and recent trends in Canadian raptors: A review." *Biological Conservation* 83:91–118.

Matthews, S. N., R. J. O'Connor, L. R. Iverson, and A. M. Prasad. 2004. *Atlas of Climate Change Effects in 150 Bird Species of the Eastern United States*. General Technical Report NE-318. U.S. Department of Agriculture, Forest Service, Northeastern Research Station, Newton Square, PA. 340 pp.

Rodenhouse, N. L., S. N. Mathews, K. P. McFarland, J. D. Lambert, L. R. Iverson, A. Prasad, T. S. Sillett, and R. T. Holmes. 2008. "Potential effects of climate change on birds of the Northeast." *Mitigation and Adaptation Strategies for Global Change* 13:517–540.

Rodriguez, C. and J. Bustamante. 2003. "The effect of weather on lesser kestrel breeding success: Can climate change explain historical population declines?" *Journal of Animal Ecology* 72:793–810.

Sanderson, F. J., P. F. Donald, D. J. Pain, I. J. Burfield, and F. P. J. von Bommel. 2006. "Long-term population declines in Afro-Palearctic migrant birds." *Biological Conservation* 131:93–105.

Sergio, F. 2003. "Relationship between laying dates of black kites *Milvus migrans* and spring temperatures in Italy: Rapid response to climate change?" *Journal of Avian Biology* 34:144–149.

# Chapter 12

# *Shorebirds*

The Arctic, where many long-distance migrant shorebirds breed, is one of the world regions most intensely affected by changing climate. The breeding habitats of many of these species are now being modified substantially. Furthermore, marine coastlines, used by many shorebirds in migration and by others for nesting, are being affected by rising sea levels and changing patterns of atmospheric and oceanic circulation. Coastal areas with high marine productivity are wintering and migratory stopover areas that host large numbers of shorebirds. Major migratory flyways of shorebirds follow the pattern of these rich stopover areas, most of which have extensive areas of mud and sand flats. The value of some of these areas is likely to be influenced by climatic change. The wintering areas of Arctic shorebirds span lower latitudes of the Northern Hemisphere, tropical coastlines, and coasts and interior wetlands of the Southern Hemisphere, all of which are experiencing climatic changes.

Many other shorebirds are short- or long-distance migrants that breed in temperate or subarctic regions and winter in more southern areas of the North Temperate Zone and tropics. These species are exposed to changing climate, as well as many direct impacts of human activity. Thus, one would expect that shorebirds would be one of the first bird groups to show serious population declines due to climate change. Migratory shorebirds thus serve as sensitive indicators of environmental change, both direct change through human modification of habitat and indirect change through anthropogenic climate alteration.

## Shorebird Populations

Populations of many shorebirds are declining (Table 12.1). Based on the International Waterbird Census by Wetlands International, nearly half of the species of the families Charadriidae and Scolopacidae show declining trends. For members of all families of shorebirds, 51 percent of intracontinental and short-distance migrants and 47 percent of intercontinental migrants are in decline. Unfortunately, for many shorebird populations in Asia and Africa, data to determine their population trends are seriously deficient.

In North America, as worldwide, many shorebird populations are in decline. Of thirty-five breeding and nonbreeding shorebirds in Canada and the United States for which adequate data are available, nineteen show significant negative population trends. These include four plovers, twelve sandpipers, and three phalaropes. An additional eight species of sandpipers and one plover show negative population trends that may not be statistically significant. Five other species show negative trends in some parts of North America, positive trends in others. Two species show overall positive trends, with only one, the Upland Sandpiper, being statistically significant. Declines are evident in populations from many different regions of North America, with multiple factors likely being contributory. Changing climate

TABLE 12.1. Status of the world's shorebirds.

| Region | Number of Species | Number of Populations | Number of Populations | | |
|---|---|---|---|---|---|
| | | | Declining | Stable | Increasing |
| North America | 42 | 86 | 31 | 20 | 6 |
| Neotropics | 56 | 109 | 25 | 22 | 4 |
| Europe | 39 | 98 | 30 | 28 | 12 |
| Asia | 65 | 198 | 31 | 16 | 7 |
| Africa | 81 | 202 | 40 | 36 | 14 |
| Oceania | 41 | 79 | 11 | 7 | 7 |
| Total | 214 | 511 | 96 | 72 | 32 |

*Source:* Data from *Wader Study Group Bulletin* 101/102, August/December 2003.

is almost certainly one of the important factors, affecting both breeding habitats and marine productivity.

The eastern North American subspecies of the Red Knot is one of the shorebirds most seriously affected. Over 20 years, its numbers have shrunk from 100,000–150,000 to 18,000–33,000. Concern has focused on its primary wintering areas in southern South America and its major spring staging area in Delaware Bay on the east coast of the United States, where birds feed heavily and deposit fat for nonstop flights of 2900 kilometers to their Arctic breeding grounds. At the Delaware Bay staging area, the birds feed primarily on eggs of spawning horseshoe crabs, which have recently become exploited heavily by commercial fisheries. Whether conditions at these wintering and staging areas are entirely responsible for decline of the population, however, is uncertain, and factors such as rising sea levels and climate change might also be involved.

Trends in numbers of other migrating shorebirds are clearly shown at Maritimes Shorebird Survey sites in eastern Canada and International Shorebird Survey sites in the midwestern and eastern United States since 1974. An analysis of these sites grouped them into two regions: coastal or near-coastal North Atlantic (eighty-one sites) and interior Midwest (fifty-four sites). For the North Atlantic region, 73 percent of the thirty shorebird species showed trends of declining abundance, and for eight of these species the trend was statistically significant. The pattern was quite different in the Midwest region, with only thirty-eight species showing declining trends, only one of which was statistically significant. Five of the eight species showing significant declines in the North Atlantic region were Arctic breeders: Black-bellied Plover, American Golden Plover, Hudsonian Godwit, Pectoral Sandpiper, and Stilt Sandpiper. Although the tundra habitat used by many of these apparently declining Arctic breeders is one of the ecosystems most intensely affected by climatic warming, it is still unclear to what extent the declines are related to climatic change.

How accurately counts of migrating shorebirds reflect their total populations is something of an issue. The most likely cause of the declines in observed numbers of migrant shorebirds in the North Atlantic region is usually assumed to be that their breeding populations have, in fact, declined. That these apparent declines are simply due to shorter stopover times by migrating birds is difficult to reject, however. Apparent declines of shorebirds at some migratory stopover sites might be the result of changed behavior of shorebirds in response to increased predation risk. In British Columbia, Canada, for example, data on Western Sandpipers and Peregrine Falcons suggest that increasing numbers of falcons have led to a reduced stopover time by sandpipers at sites most prone to predation. Although some workers have concluded that reduced stopover time may be contributing to an apparent reduction of the population of shorebirds in migration, it appears unlikely that these relationships can account for apparent declines of so many diverse species.

Many European shorebird populations are also in decline. One study examined population trends in fourteen species of European long-distance migrant shorebirds for the periods 1970–1990 and 1990–2000, using the Birds in Europe database. During the 1970–1990 period, only three species showed significant trends of decreasing population: Collared Pratincole, Great Snipe, and Wood Sandpiper. Populations of the Whimbrel and Green Sandpiper, on the other hand, increased significantly. During the decade of the 1990s, five species showed significant declines: Black-winged Pratincole, Ruff, Great Snipe, Marsh Sandpiper, and Common Sandpiper. The Black-winged Stilt, Common Greenshank, and Green Sandpiper increased in numbers. For the Black-winged Stilt, Great Snipe, and Whimbrel, changes appeared collectively to be greater than for congeneric or closely related nonmigratory species. A rapid decline has also occurred in the Temminck's Stint population wintering in the Baltic Sea region. The cause of the decline of this Arctic breeder is uncertain but seems to involve reduced adult survival.

On the other hand, populations of some northern species are doing well. Icelandic populations of Black-tailed Godwits, which winter in England, are increasing, for example. These birds are expanding their use of both breeding and wintering habitats.

Although coastal marine habitats farther south in the North Temperate Zone are certain to be affected by rising sea levels, little evidence exists to date of negative impacts on shorebird populations. In England, between 1985 and 1996, breeding populations of the Common Redshank declined by 23 percent in salt marshes subject to the effects of gradually rising sea level. The primary cause of this decline, however, appears to be increased livestock grazing in salt marshes. A modeling study of the habitat relationships of breeding Eurasian Oystercatchers in the Wadden Sea, in the Netherlands, suggested that rising sea levels are likely to cause few problems. On the other hand, wetlands along the Mediterranean coastline are predicted to be severely threatened by rising sea level by AD 2080, so birds of some coastal areas are likely to be affected badly.

Along the East Asian–Australasian Flyway, populations of most species seem to be stable. A few species, however, including the Spoon-billed Sandpiper, Curlew Sandpiper, and Red-necked Stint, appear to be declining.

These species all breed in Arctic Siberia. The Spoon-billed Sandpiper is estimated to have a total population of less than 3000 individuals.

Little information is available on shorebird populations in the Southern Hemisphere. In eastern Australia, expansion of irrigated agriculture is diverting water from many interior wetlands heavily used by both resident and wintering shorebirds. These developments alone are probably responsible for the 81 percent decline in populations of resident shorebirds between the 1980s and 2006. In addition, however, wintering migratory shorebirds have declined in numbers by 73 percent over the same period. For these species, loss of wintering habitat, loss of staging and stopover habitat along the East Asian–Australasian Flyway, and Arctic warming all are likely to be factors.

## Breeding Ranges

Some fifty or so species of shorebirds depend on breeding sites in Arctic tundra areas of northern North America and Eurasia. For these species, Arctic warming may be beneficial in the short term, by creating longer breeding seasons and increasing the abundance of insect foods. In the longer term, vegetation changes are certain to reduce the area of suitable breeding habitat for shorebirds, essentially compressing the area of suitable tundra between forested regions and the Arctic Ocean.

Climate models show how extensively the breeding ranges of many of these shorebirds are at risk. One global circulation model predicts that an atmospheric doubling of carbon dioxide would lead to climate change that would ultimately reduce the area of Arctic tundra by 40–57 percent. Most of the lost habitat would represent tundra invaded by boreal forest. Only very limited expansion of tundra is projected to occur, largely in Greenland, because of warming of the Arctic. The specific habitats of several species of tundra-breeding sandpipers were predicted to decline by 5 to 57 percent. One of these species, the Spoon-billed Sandpiper, now listed as threatened, would lose an estimated 57 percent of its breeding habitat.

Another modeling study predicts the loss of breeding habitat by seven sandpiper species in Finland and adjacent portions of Norway, Sweden, and Russia (Table 12.2). Two scenarios were examined, one assuming global warming of 2.0°C by AD 2100, the second a warming of 3.8°C. Under both scenarios, the Jack Snipe, Bar-tailed Godwit, and Spotted Redshank were predicted to lose over 90 percent of breeding habitat by AD 2051–2080. The Broad-billed Sandpiper and Red-necked Phalarope were

TABLE 12.2. Percent declines in breeding habitat area of selected shorebirds as predicted by the general circulation model HadCM3 for northern Scandinavia under global warming scenarios B1 (2.0°C by AD 2100) and A2 (3.8°C by AD 2100).

| Model Scenario | B1 (AD 2051–2080) | A2 (AD 2021–2050) | A2 (AD 2051–2080) |
|---|---|---|---|
| Lowland Tundra Species | | | |
| Jack Snipe | −94.8 | −66.1 | −97.7 |
| Bar-tailed Godwit | −98.3 | −84.2 | −99.6 |
| Whimbrel | −56.5 | −35.0 | −70.5 |
| Spotted Redshank | −97.5 | −73.0 | −99.1 |
| Common Greenshank | −43.2 | −37.4 | −67.9 |
| Broad-billed Sandpiper | −75.0 | −41.7 | −87.8 |
| Red-necked Phalarope | −61.0 | −42.8 | −80.7 |
| Montane Tundra Species | | | |
| Eurasian Dotterel | −89.5 | −86.2 | −98.9 |
| Temminck's Stint | −68.0 | −51.4 | −82.0 |

Source: Data from Virkkala et al. 2008.

projected to lose over 80 percent of habitat under the warmer scenario and over 60 percent under the milder scenario. The Whimbrel and Common Greenshank were projected to lose less breeding habitat, but still nearly 70 percent under the warmer scenario. Species nesting in mountain heathland will also be affected severely. The Eurasian Dotterel will likely lose almost all breeding habitat by AD 2051–2080 under these warming scenarios, and Temminck's Stint 68–82 percent.

Some shorebirds are expanding or contracting their breeding ranges. In North America, the American Golden Plover has increased in numbers at Devon Island in the high Arctic. Wilson's Phalarope appears to have expanded its range northward and westward into the Yukon Territory of Canada and southeastern Alaska. The abundance of Red and Red-necked Phalaropes in southern parts of their North American breeding range, on the other hand, seems to be decreasing.

In Siberia and northern Greenland, several shorebirds have expanded their ranges or established breeding populations in new locations correlated with climatic amelioration. In Finland, most shorebirds with distributions in more-southern parts of the country showed a northward expansion of their breeding ranges between 1974 and 1989. These included the Little Ringed Plover, Eurasian Woodcock, Green Sandpiper, and Eurasian Curlew. The more-northern Whimbrel and Broad-billed Sandpiper also showed northward range expansions. Several of the shorebirds breeding farthest north, including the Jack Snipe and Spotted Redshank, saw their ranges retreat southward.

The Ruff, a Eurasian shorebird that breeds both in Arctic tundra habitats and in wet grassland habitats farther south, has declined greatly in its southern breeding localities. It has become extirpated in former breeding localities in Britain, France, and Belgium, and is declining in most localities from the Netherlands and Germany east into Russia. The decline of the species in many areas appears to have occurred in spite of protection of its habitat. Many of the wet grassland areas in the southern part of the Ruff's range are affected by climate warming and eutrophication by nitrogen enrichment due to rainfall deposition and fertilizer runoff. The dense wetland vegetation appears to be less suitable to the Ruff. Thus, the species is being forced to shift its range northward and eastward, probably at least in part because of climate change. Other marsh-nesting shorebirds, including the Common Snipe, Common Redshank, and Black-tailed Godwit, may also be affected by this same pattern of habitat change.

## Migration Schedules and Stopover Patterns

Arctic-breeding shorebirds appear to have migration routes and schedules that are very conservative. Many Arctic shorebirds arrive at breeding areas via long-distance flights from staging areas in the North Temperate Zone, often crossing wide ocean areas. At departure areas for these flights, weather conditions are only weakly correlated with those of breeding areas. Changes in wind patterns and weather conditions along migration routes could also be substantial challenges for these species.

For long-distance migrants, staging and stopover areas are clearly of critical importance. Worldwide, about 53 coastal marine areas that host more than 100,000 shorebirds in spring and fall migration have been identified. Some of these, such as Delaware Bay in North America, the Wadden Sea in Europe, and Saemangeum, South Korea, in particular, have suffered from fisheries overexploitation, pollution, or development. How these migration sites will respond to rising sea level is difficult to predict, as both ocean level rise and sediment deposition must be considered. Along the

Atlantic coast of North America, even if sea level rise exceeds sediment deposition, the result may benefit migratory shorebirds by increasing the areas of bare mud flats and shallow water where they forage.

Changes in habitats used for migration stopovers, due either to climate change or human disturbance, already appear to be contributing to declines of some shorebirds. In North America, the shorebird populations most in decline are those following continental migration routes, rather than oceanic routes. Large-scale habitat change, including intensified monocultural farming patterns and wetland degradation, seem largely to be responsible for these declines. Increasing aridity in the North American interior would obviously exacerbate this trend.

Worldwide, spring migration schedules of many shorebirds are earlier in some locations but little changed in others. In some Arctic areas, such as Iceland and northern Greenland, where birds arrive by long flights from stopover areas far to the south, little advance in arrival is seen. In the Yukon–Kuskokwim Delta of Alaska and the Bering Strait area of Russia, for example, long-distance migrant shorebirds showed no trend of earlier arrival between the late 1970s and early 2000s, in spite of substantially warmer conditions in early summer.

In other Arctic areas of the Old World, at least some species show advances of spring arrivals at Arctic and subarctic breeding areas. On the Kola Peninsula, Russia, Common Sandpiper and Common Snipe advanced their arrival dates 6–7 days over the period 1931–1999. At Tromsö, Norway, the Eurasian Golden Plover advanced its arrival a little over 7 days during the period 1978–2000. Several other species, including the Eurasian Oystercatcher, Northern Lapwing, and Redshank, also showed arrival patterns that were earlier in warmer years, suggesting that if consistent patterns of warming are established in these areas, many species will respond by arriving earlier. Arrival dates of Black-tailed Godwits in Iceland are positively correlated with the March value of the North Atlantic Oscillation (NAO).

Spring arrivals have advanced most for partial or short-distance migrants in more southern areas. In Scotland, the Eurasian Curlew showed an advance in arrival date of over a month between 1974 and 2000. In Oxfordshire, England, the Little Ringed Plover arrived over 3 weeks earlier in 2000 than in 1971. Departure in the fall has also advanced by about the same period, so time spent on the breeding range has remained about the same. In Lithuania, the Northern Lapwing had advanced its arrival by more than a week during the 1990s, compared with the 1970s and 1980s. In 2008 avocets returned to the Wildfowl and Wetlands Trust's reserve at

Martin Mere in the UK at least 3 weeks earlier that usual. This is the earliest Lancashire record for the return of wading birds for breeding. In the southern Ural Mountains of Russia, the Northern Lapwing advanced its spring arrival at the Il'men' Reserve by almost 5 days over the period 1971–2005.

Migrants on spring passage have also shown some advances in timing in Europe. Data on migratory passage of Spotted Redshanks, Greenshanks, and Wood Sandpipers were obtained at four central European localities between the late 1960s or mid-1970s to 2002. All three species showed advances of spring migration and delays for autumn migration. Weather variables seem to account for much of the advance in spring passage, with warm spring conditions and positive conditions of the NAO favoring early passage. Weather on the breeding ranges was largely responsible for change in timing of fall migration, with cool spring weather and late snowmelt tending to result in delayed fall migration. These patterns indicate that the species have considerable ability to adjust their migration schedules in response to weather encountered during migration. Farther east, the dates of spring passage of many of the shorebird migrants at Jurmo Island and the Hanko Peninsula of Finland are significantly earlier during warmer phases of the NAO.

At lower latitudes in North America, many species show earlier spring arrivals and passage dates. In Worcester, Massachusetts, both short-distance migrants such as the Killdeer, Spotted Sandpiper, Common Snipe, and American Woodcock and long-distance migrants such as the Solitary Sandpiper and Pectoral Sandpiper have advanced their spring arrivals. Between 1932 and 1993, the arrival of these species became about 2–3 weeks earlier. In the Cayuga Lake Basin of New York State, the Killdeer, Common Snipe, American Woodcock, Semipalmated Sandpiper, and Least Sandpiper were also arriving earlier in 1951–1993 compared with 1903–1950. American Woodcocks arrived in Maine about 8 days earlier in 1994–1997 than in 1899–1911, although the arrival dates of Spotted Sandpiper and Common Snipe did not change over this same period.

Farther west, arrival of some short-distance migrant shorebirds at Delta Marsh, Manitoba, Canada, became significantly earlier over the period 1939–2001. The Killdeer advanced its arrival by about 7 days and the Willet by about 10 days. On the other hand, the Greater Yellowlegs and Lesser Yellowlegs tended to arrive about 11 and 22 days later, respectively, at this staging area, which is only a short distance south of their breeding range. Several other resident or migrant shorebirds showed no significant arrival trend.

## Breeding Schedules

Few data exist on changes in nesting schedules of shorebirds, especially in Arctic breeding areas, where substantial climate change has occurred and the need for adjustment of nesting schedules is great. Nevertheless, it is clear that the phenology of insect food abundance has changed in many locations. On the Taimyr Peninsula in Siberian Russia, for example, the insect abundance peak advanced by 7 days over the period 1973–2003. On the southern Taimyr Peninsula, some advance of egg-laying dates by shorebirds has occurred, although this has not been noted in Alaska.

Farther south, significant advance in breeding has occurred for some species. A remarkable set of data is available on the date of first egg laying by the Northern Lapwing in the Netherlands. These data cover the period 1897 to 2003. Since the 1950s, the date of first egg laying has advanced about 10 days, on average. Even more striking is the relationship between laying date and average temperature from mid-February to mid-March. The warmer the weather, the earlier the birds begin laying. In Britain the laying dates for the Eurasian Oystercatcher, Eurasian Curlew, and Redshank advanced over the period from 1971 to 1995.

## Wintering Habitats and Schedules

Climatic change is leading to substantial changes in some shorebird wintering areas. Several wintering shorebirds have shifted their wintering areas from the west coast of Britain toward eastern and southern coasts as winters have become milder. Several common shorebirds in Britain and western Europe have shown changes in wintering patterns over a 30-year period beginning in the late 1970s. The Eurasian Oystercatcher, Red Knot, Black-bellied Plover, Dunlin, Eurasian Curlew, Bar-tailed Godwit, and Common Redshank have all shown appreciable shifts in the center of their wintering area, statistically significant except for the Eurasian Oystercatcher. These shifts were eastward or northeastward except for the Redshank, which moved northwestward. In general, wintering ranges moved toward the colder margins of their wintering areas, indicating that formerly less favorable wintering sites were being used by greater numbers of birds. Most of these shorebirds come from breeding ranges in northern Eurasia, except for many Redshanks that breed in the Faeroe Islands and Iceland, northwest of their wintering area. Thus, wintering range shifts, in general, appear to be in the direction of breeding areas. Range shifts also showed a strong positive correlation with mean January temperatures, especially at the colder edges of the wintering range. In other words, shifts were toward areas with warmer winters. The Common Ringed Plover has also shifted much of its wintering population from the west coast to the east coast of Britain. Continued winter warming appears likely to promote additional range changes.

The possible impacts of rising sea level due to global warming were examined for shorebirds wintering in two estuaries in England, one on the Irish Sea and the other on the North Sea. Five species that use open mud or sand flats for foraging were considered: Eurasian Oystercatcher, Red Knot, Dunlin, Eurasian Curlew, and Common Redshank. Models of estuary shape in relation to sea level rise were combined with models of bird population densities in relation to estuarine conditions. Four scenarios of sea level rise and three different societal responses to protection of land against rising sea level were considered. The maximum sea level rise by the mid-2000s was estimated at 67–78 centimeters, depending on land level change. For the two estuaries, little change is predicted in the area of estuarine habitat unless current defenses against seawater intrusion are breached or moved to more inland locations. Where rising sea level does expand the area of estuarine conditions, the result is predicted to be the creation of sandier intertidal habitats, favoring the Eurasian Oystercatcher, Red Knot, and Eurasian Curlew. The greater extent of intertidal sand and mud flats, however, is likely to increase the total capacity of the estuaries to support shorebirds in general. Other coastal habitats used by shorebirds may not be modified in the same manner, however, and habitat loss due to rising sea level is of serious concern.

The timing of wintering shorebird arrivals and departures has also changed in some instances. In Essex, UK, the Jack Snipe has delayed its fall arrival by about 6 days per decade since 1966–1974. In southeastern Australia, between 1985 and 2003, wintering Red Knots advanced their arrival by almost 23 days, and wintering Eurasian Curlews advanced slightly over 8 days between 1978 and 2004. Both of these species breed in Siberia, and their earlier arrival in Australia suggests a correspondingly earlier departure from the breeding range. On the other hand, the Double-banded Plover, which breeds in New Zealand and the Chatham Islands, delayed its arrival by somewhat over 12 days between 1977 and 2006, suggesting that conditions in the breeding range were remaining favorable for a longer time.

# Summary

Shorebirds are sensitive indicators of global climate change, and the populations of many short-distance and long-distance migratory species are in decline. Although climatic warming may seem to favor species breeding in the Arctic, a mismatch may develop between the time of nesting and the time of peak food availability. Ultimately, vegetation change may cause extensive loss of breeding habitat. Breeding ranges of some shorebirds have shifted northward, and spring arrival on breeding areas has advanced for some species. The skimpy information available on migration to wintering areas suggests that some species have advanced and some have delayed their arrivals. Rising sea levels are certain to affect migrating and wintering shorebirds, perhaps favoring some and disfavoring others.

## KEY REFERENCES

Austin, G. E. and M. M. Rehfisch. 2005. "Shifting nonbreeding distributions of migratory fauna in relation to climate change." *Global Change Biology* 11:31–38.

Bart, J., S. Brown, B. Harrington, and R. I. G. Morrison. 2007. "Survey trends of North American shorebirds: Population declines or shifting distributions?" *Journal of Avian Biology* 38:73–82.

Erwin, M. E., G. M. Saunders, D. J. Prosser, and D. R. Cahoon. 2006. "High tides and rising seas: Potential effects on estuarine waterbirds." *Studies in Avian Biology* No. 32:214–228.

International Wader Study Group. 2003. *Waders Are Declining Worldwide: Conclusions from the 2003 International Wader Study Group Conference, Cádiz, Spain.* Wader Study Group Bulletin 101/102:8–12.

Maclean, I. M. D., G. E. Austin, M. M. Rehfisch, J. Blew, O. Crowe, S. Delanys, K. Devos, et al. 2008. "Climate causes rapid changes in the distribution and site abundance of birds in winter." *Global Change Biology* 14:2489–2500.

Meltofte, H., T. Piersma, H. Boyd, B. McCaffery, B. Ganter, V. V. Golovnyuk, K. Graham, et al. 2007. "Effects of climate variation on the breeding ecology of Arctic shorebirds." *Meddelelser om Grønland—Bioscience* 59. Danish Polar Center, Copenhagen. 48 pp.

Morrison, R. I. G., Y. Augrey, R. W. Butler, G. W. Beyersbergen, G. M. Donaldson, C. L. Gratto-Trevor, P. W. Hicklin, V. H. Johnson, and R. K. Ross. 2001. "Declines in North American shorebird populations." *Wader Study Group Bulletin* 94:34–38.

Nebel, S., J. L. Porter, and R. T. Kingsford. 2008. "Long-term trends of shorebird populations in eastern Australia and impacts of freshwater extraction." *Biological Conservation* 141:971–980.

Piersma, T. and Å. Lindström. 2004. "Migrating shorebirds as integrative sentinels of global environmental change." *Ibis* 146:61–69.

Rehfisch, M. M., G. E. Austin, S. N. Freeman, M. J. S. Armitage, and N. H. K. Burton. 2004. "The possible impact of climate change on the future distributions and numbers of waders on Britain's non-estuarine coast." *Ibis* 146(Suppl. 1):70–81.

Thomas, G. H., R. B. Lanctot, and T. Székely. 2006. "Can intrinsic factors explain population declines in North American breeding shorebirds? A comparative analysis." *Animal Conservation* 9:252–258.

Virkkala, R., R. K. Keikkinen, N. Leikola, and M. Luoto. 2008. "Projected large-scale range reductions of northern-boreal land bird species due to climate change." *Biological Conservation* 141:1343–1353.

Zockler, C., S. Delany, and W. Hagemeijer. 2003. "Wader populations are declining: How will we elucidate the reasons?" *Wader Study Group Bulletin* 100:202–211.

# Chapter 13

## *Waterfowl and Other Waterbirds*

Waterfowl, comprising ducks, geese, and swans, are especially vulnerable to changing climate, largely through alteration of their breeding habitat. With global climate change, much of the area of freshwater habitat used by nesting ducks is projected to become subject to increased drought. Many species of ducks, geese, and swans that utilize coastal or tundra breeding areas are also projected to be affected by rising sea levels and vegetation change due to climatic warming.

Other large migratory waterbirds include loons, grebes, rails and their relatives, storks, cranes, herons and egrets, ibis, pelicans, and gulls and terns. Climatic changes that affect both interior and coastal wetlands may be significant for many of these species. On the other hand, for many waterfowl and other large waterbirds, the specific patterns of migration appear to be culturally transmitted, and changes in breeding and nonbreeding ranges, migration routes, and timing of initiation of breeding may occur quickly in response to changing habitat conditions.

## North America

In North America, both waterfowl and many other wetland birds have shown changes in populations and migration patterns. Projected climatic changes are likely also to affect many wetland habitats.

## Ducks, Geese, and Swans

Climatic change will certainly alter breeding habitats for waterfowl in North America in diverse ways. The numbers of pond habitats for breeding ducks in central North America have always fluctuated enormously, from lows of slightly over 3 million to highs of over 8 million. Since 2002, pond numbers had tended to increase, in 2007 reaching 7 million. In 2008, however, the number of ponds decreased to 4.4 million. Total duck numbers have varied greatly, as well, from a low of slightly over 25 million in 1990 to a high of over 43 million in 1999. Since 1999, numbers declined to 31.7 million in 2005, recovered to 41.2 million in 2007, and declined again to 37.3 million in 2008. Climatic change will alter the extent of breeding habitats, as well as intensify the year-to-year variability in habitat favorability.

Climatic change has already affected some species of ducks. Populations of the American Black Duck and the Northern Pintail breeding in eastern Canada and the northeastern United States have declined in numbers almost continuously since the 1950s. Several factors are probably involved in the decline of Northern Pintail populations, including drought and agricultural expansion in the Canadian Prairie Provinces. Populations of Greater and Lesser Scaups have also declined since the mid-1980s in northwestern boreal forest areas, the northern Rocky Mountains, and the western Dakotas. Although several factors may contribute to Lesser Scaup declines, much of their breeding range has already experienced widespread climatic warming, and this change appears likely to be involved to some degree. In Alaska, where the boreal forest breeding area for Lesser Scaup has warmed by about 3.0°C since the 1960s, for example, populations of this duck have declined substantially. Climate-induced changes in the water chemistry of breeding ponds seem to be favoring limnetic food chains as opposed to the benthic food chains that support invertebrates on which both adult and young scaup feed. On the other hand, populations of these ducks in the Prairie–Parkland Region of Canada have tended to increase.

In the northern Great Plains of North America, the so-called Prairie Pothole Region, two general circulation models of future climate, based on a doubling of atmospheric $CO_2$, predict major losses of waterfowl habitat (Figure 13.1). By AD 2060, these models suggest, in the region stretching from northern Iowa to central Alberta, the number of breeding ponds will decline from the present 1.3 million to about 0.6–0.8 million. This corresponds to a reduction in duck populations in this region from about 5.0 million to 2.1–2.7 million. The species to be affected heavily include the Mallard, Northern Pintail, Northern Shoveler, Gadwall, Blue-winged Teal,

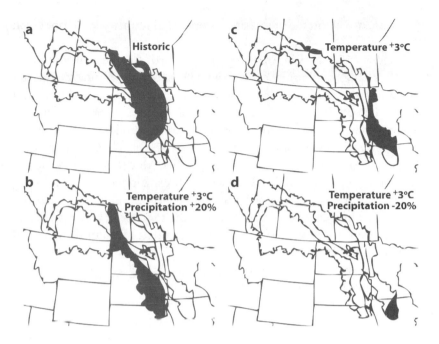

FIGURE 13.1. Simulated shift in region of optimal conditions for waterfowl breeding in the northern plains region of the United States and Canada under three scenarios of climatic warming and change in precipitation. (From Johnson, W. C. et al. 2005. *Vulnerability of northern prairie wetlands to climate change*. Copyright 2005, American Institute of Biological Sciences.)

Canvasback, Redhead, Lesser Scaup, and Ruddy Duck. Earlier studies in the Canadian Prairie Provinces showed that populations of ducks are very sensitive to climatic shifts. Populations of ten species of ducks declined in most parts of the Canadian Prairie–Parkland Region between 1955 and 1989. Climatic variation was one of the most important factors in determining their abundance. In addition, however, expansion of agriculture contributed to the decline of populations, especially in the drier western prairies.

Models that combine climate, vegetation, and bird relationships in the northeastern United States are being used to predict responses of bird species on Breeding Bird Survey (BBS) routes to climate change during the twenty-first century. These models project changes in the fraction of routes with particular species. Based on the model that assumes high emission levels of greenhouse gases, the occurrences of Mallard ducks and Canada Geese are predicted to decline by about 38–40 percent. The occurrence of

Blue-winged Teal, on the other hand, is projected to increase by more than 200 percent.

Waterfowl habitats in Arctic regions of North America will also be affected by changing climate. Sea level rise will likely cause habitat loss in Arctic coastal plain regions of Alaska and Canada, which are important nesting areas for geese, sea ducks, and Tundra Swans. On the other hand, climatic warming may expand nesting habitat for some species. On the Arctic coastal plain of Alaska, for example, two duck species of conservation concern are the Spectacled Eider and Steller's Eider. Modeling studies suggest that warming of 4–6°C by AD 2040 will likely expand the potential nesting habitat of both species. Potential nesting habitat for the Brant will also increase. These projections are coupled with the caveat that extensive petroleum development in the coastal region could offset much of the potential benefit of increased habitat.

In contrast to some ducks, most populations of geese in North America were stable or on the increase from 1998 to 2007. One of these populations, the Aleutian Islands race of the Canada Goose, fell to about 800 birds in 1967 and was listed as endangered. This population increased to about 114,000 birds in 2007–2008. Tundra Swans, especially the western population, have generally increased in numbers since the 1970s.

Migration patterns of many waterfowl are changing under the influence of global warming. At the Delta Marsh in Manitoba, Canada, where data on spring arrival of waterfowl are available for the period 1939–2001, temperature trends for February through May suggest a gradual warming, although year-to-year variability is great. Snow Geese, Canada Geese, and eight species of dabbling and diving ducks showed significant tendencies to arrive earlier during this period. All 16 species of short-distance migrant ducks showed significant trends toward earlier arrival in warmer springs. The single long-distance migrant, the Blue-winged Teal, also showed a significant trend toward earlier arrival over the 62-year period, as well as significantly earlier arrival in warmer springs. Two-thirds of the waterfowl species breeding at the Delta Marsh significantly advanced their arrival dates.

Migration schedules for at least waterfowl have also become earlier in most of the United States. Canada Geese advanced their arrival in southern Wisconsin by almost a month over the period from 1936 to 1998. In New York and Massachusetts, Blue-winged Teal showed a significant tendency for earlier arrival in spring during the twentieth century. In Massachusetts, Blue-winged Teal arrival was over a month earlier in the latter half of the twentieth century than in the first half. Over the period 1970–

2002, Wood Ducks advanced their spring arrival at Middleborough, Massachusetts, by over a month. In Maine, however, this species showed no significant change in arrival date in the late 1990s compared with the early 1900s.

Nesting times of some northern geese have also advanced. Snow and Canada Geese in the Hudson Bay area of Canada advanced their nesting by about 30 days between 1951 and 1986. At Bylot Island in the high Canadian Arctic, on the other hand, no advance in initiation of nesting by Snow Geese was detected between 1989 and 2004, although warming of the area's climate by about 1.8°C occurred. Temperatures there are predicted to rise by about 4°C by AD 2090, and an advance of nesting by 5 days is predicted.

Deterioration of staging and stopover areas for migrating waterfowl due to climatic warming is also a concern. In coastal areas such as the Bay of Fundy and Chesapeake Bay, rising sea levels may reduce the carrying capacity of marshlands essential for preparing birds for migration to wintering and breeding grounds. In southern Manitoba, drought conditions are already known to impair the accumulation of body reserves by Snow Geese on their way to Arctic breeding grounds.

Wintering areas will also be affected by climate change. Sea level rise will lead to loss of coastal wetlands from the mid-Atlantic region to the Gulf of Mexico coast, where many waterfowl winter. In areas with seawalls that protect developed inland areas, the loss of wetlands is likely to be 2–45 percent, and in unprotected coastlines, 17–43 percent. In Chesapeake Bay, for example, sea level rise of 43 to 74 centimeters during this century is predicted to reduce the extent of tidal marshes and eelgrass beds that support major wintering populations of several ducks, including Redhead, Canvasback, Northern Pintail, American Widgeon, Black Duck, and Ruddy Duck. Several of these species have declined markedly in wintering numbers since the 1950s because of the declines of submerged aquatic vegetation that have already occurred. On the Pacific coast, sea level rise and reduced inflow of freshwater is likely to affect estuary areas where waterfowl winter, with severe impacts on species of diving ducks such as Canvasbacks and Ruddy Ducks. In the western United States, areas such as the Klamath Basin of California and Oregon, winter home to about 6 million ducks, geese, and swans, is likely to be affected by reduced freshwater inflow and increased diversion for irrigation.

On the other hand, climatic warming may improve winter habitats in more northern areas. Many waterfowl species appear to be shifting their winter ranges, according to data from Audubon Society Christmas Bird

Counts over the past 40 years. About seventeen species of ducks, geese, and swans showed significant northward expansions averaging about 209 kilometers. On the other hand, another sixteen species showed significant southward shifts averaging about 148 kilometers. Several of the species showing southward expansions were birds wintering in coastal marine waters. Observations in a number of local areas give similar patterns. The number of wetland birds, both freshwater and marine, wintering on Cape Cod, Massachusetts, for example, nearly doubled from 1930 through 2001.

## Other Waterbirds

The population dynamics of Whooping Cranes breeding in Wood Buffalo National Park, Northwest Territory, Canada, and wintering at Aransas National Wildlife Refuge, Texas, are closely related to the Pacific Decadal Oscillation (PDO). Growth, with a lag time of 1 year, shows a direct relation to the PDO Index, for which positive values represent the warm phase. Since 1977, the PDO has tended to be in its warm phase, except for 1989–1991 and 1999–2001, and the crane population grew from about 72 individuals to 217. Cold phase PDO conditions led to poor reproductive success at Wood Buffalo National Park. Warm phase conditions favored successful reproduction, except when the warmth was associated with drought in the breeding range. As we noted earlier (Chapter 2), the persistence of the warm phase of the PDO may or may not represent an influence of global warming.

In the Arctic, some waterbirds may be suffering from alteration of food chains due to global warming. In Arctic regions of Alaska, the Red-throated Loon declined in numbers by almost 50 percent between 1977 and 2004. This species nests on freshwater ponds but forages in the ocean during the nesting period. Nesting birds were usually unable to raise more than one chick during the study in 2002–2003, apparently due to deficiency of high-quality fish prey in the Bering Sea. The decline in loon numbers appears to be related to the regime shift from cold to warmer water in the Bering Sea, beginning in the late 1970s.

Farther south, other waterbirds may eventually suffer from climatic warming. The modeling analysis of climate, vegetation, and bird incidence on BBS routes in the northeastern United States, mentioned earlier, also projects that a few species of other waterbirds will see declines. These include the Common Loon, American Bittern, and Sora Rail. On the other hand, populations of the White Ibis and most herons and egrets are likely to increase substantially.

In coastal areas, rising sea levels will affect many birds that nest in coastal marshes with emergent vegetation. Rising sea level is likely to reduce the area of vegetated marshland that many nesting seabirds utilize. About thirty-three species of waterbirds breed in mid-Atlantic coastal marshes, including several declining species of special conservation concern, such as the Black Rail, Black Skimmer, American Oystercatcher, and a number of terns. Many more species use these marshes in migration and during the winter, as well.

Migration patterns of some of these other waterbirds have changed. The Virginia Rail and the Green Heron began arriving significantly earlier in Worcester County, Massachusetts, over the period 1932–1993. In the Cayuga Basin, New York, both the Virginia Rail and the Sora Rail had advanced their spring arrivals significantly during the last half of the twentieth century, compared with the first half. In southern Wisconsin, the Great Blue Heron had advanced its spring arrival by more than 11 days over the 61-year period ending in 1998. In Maine, this species arrived about 10 days earlier in 1994–1997 compared with 1899–1911.

Arrival dates of a number of other waterbirds at Delta Marsh in Manitoba, Canada, also advanced significantly during the period 1939–2001. These included Western Grebes, American White Pelicans, and Sandhill Cranes. No waterbird had a significantly delayed arrival pattern during the 62-year period.

Wintering ranges of many waterbirds have shown latitudinal shifts, as indicated by occurrences in Audubon Christmas Bird Counts over the past 40 years. Some gulls, such as the Ring-billed Gull, have begun wintering much farther north, while others, particularly Bonaparte's Gull, have shifted their ranges southward. Five species of rails, including species using freshwater and saltwater marsh habitats, showed significant northward expansions. Seven herons and egrets also showed northward range expansions, whereas only three showed a slight tendency to retreat southward. Other waterbirds showed mixed responses.

## Europe

In western Eurasia, waterfowl and other waterbirds have shown changes in ranges, populations, and migration patterns similar to those in North America.

## Ducks, Geese, and Swans

Waterfowl ranges and populations have received less attention in Europe than in North America. In Finland, between 1974 and 1989, the Gadwall, a more southern species in the country, expanded its breeding range northward substantially. Data from the Birds in Europe database show that the Garganey, a long-distance migrant wintering largely in sub-Saharan Africa, declined in abundance throughout much of Europe during the period 1970–2000. In Finland, the breeding range of the Whooper Swan has retreated southward about 200 kilometers.

The ranges of some geese are expanding northward, and their populations are growing. Barnacle Geese and Brant have also expanded their breeding areas northward to the Kanin Peninsula in northern Russia. In Norway, Barnacle Geese have expanded their use of spring staging areas to more-northern islands and to islands closer to the mainland with more extensive cropland. This shift is probably in part due to intensive grazing on the outer islands, combined with their rapidly growing total populations. Increasing spring temperatures may also have contributed to the northward shift of staging areas. Populations of Brant, Barnacle Geese, and Pink-footed Geese on the Arctic Svalbard Islands have all increased in recent decades, to the point that they may be exceeding the carrying capacity of breeding habitat. The Svalbard Pink-footed Goose population has more than tripled in size since the 1960s. Climatic warming might benefit these populations. Modeling approaches suggest that with a warming of summer temperatures by 1.0°C, the area of habitat suitable for nesting by Pink-footed Geese on the Svalbard Islands would increase by 84 percent, and with warming of 2.0°C it would more than double.

In spite of these patterns of population increase, modeling studies suggest that future changes in some tundra goose populations may be quite different. In Eurasian tundra regions, populations of geese appear likely to decline substantially with the warming associated with a doubling of $CO_2$. A general circulation model used by one group of investigators (HadCM2GSa1) indicates that the 8.4–10.4 million geese of ten species are likely to decline to about 4–5 million as tundra breeding habitats are displaced by forest and taiga. This decline is particularly serious for tundra populations of three Eurasian species considered to be globally threatened; those of the Bean Goose are projected to decline by 76 percent, those of the Red-breasted Goose by 67 percent, and those of the Emperor Goose by 54 percent.

Migratory waterfowl in Europe also have received little attention with regard to changing migration patterns. On Jurmo Island and the Hanko

Peninsula of southern Finland, fourteen species of ducks and geese showed strong trends for earlier passage during warm phases of the North Atlantic Oscillation (NAO). The Pink-footed Goose, which breeds in Arctic Greenland, Iceland, and Svalbard and winters in Europe, is advancing its migration somewhat. The population of Pink-footed Geese that winter in Denmark and nest in Svalbard advanced their spring departure from Denmark by about 2 weeks between 1990 and 2004. Arrival of these birds at their final migratory stopover area in northern Norway had previously advanced by more than a week between 1981 and 1991.

Barnacle Geese breeding on the Barents Sea coast of Russia advanced their departure dates from the Wadden Sea in the Netherlands by about 10 days during the late 1900s, corresponding to about 10 days per degree Celsius of spring warming. At stopover sites in the Baltic Sea, departure has advanced only about 2 days, and at their breeding area at the Pechora Delta in Russia, no advance in arrival was detected.

In Russia, a few data are available on other waterfowl migration schedules. On the Kola Peninsula, Lapland, the Whooper Swan and the Green-winged Teal showed no advance in arrival over the period 1931–1999, but the Common Goldeneye and Mallard were arriving 1–2 weeks earlier. The advances in arrival appear to be correlated with warmer spring weather during the 1970–1999 period. In the fall, Whooper Swans have tended to depart about 12 days earlier, correlated with earlier lake freezing. In the southern Ural Mountains of Russia, the Garganay appears to have delayed its spring arrival at the Il'men' Reserve by almost 11 days over the period 1971–2005.

With regard to wintering relationships, even less information is available. Since 1966–1974, Tundra Swans are arriving 7–18 days earlier per decade to their wintering areas in Essex, UK. At the Doñana wetlands in southern Spain, wintering populations of ducks fluctuated greatly in numbers from 1978 to 2005. Since the early 1990s, several species have declined in numbers, including the Gadwall, Green-winged Teal, Shoveler, and Common Pochard. Many of these species are long-distance migrants breeding in northern Europe. Whether overall populations of these species are declining or wintering areas have shifted farther north is uncertain. The Red-crested Pochard, a resident in Spain, has tended to increase.

## Other Waterbirds

Ranges and populations of several European waterbirds have changed in apparent response to changing climate. In Finland, a number of waterbirds with southern distributions, including the Great Crested Grebe, Gray Heron, Water Rail, and Eurasian Coot, have expanded their ranges northward. The Moorhen, on the other hand, retreated southward. Across Europe, several species of herons, storks, and ibis showed significant population declines during the period 1970–1990, but these did not continue through the 1990s. For the Black Stork and the White Stork, populations increased during this most recent decade. A similar pattern of strong population decline was seen for several species of crakes during the period 1970–1990, with declines lessening during the 1990s.

Many large waterbirds have shown changes in their migration schedules. White Storks throughout Europe have advanced their spring migrations. In Spain, this species advanced its arrival by about 40 days over the period 1944–2004. Most of this change occurred during the last 20 years of this period. Earlier arrival has also resulted in a longer period of summer residency. In western Poland, first-arrival dates of the White Stork have advanced about 10 days over their arrival dates a century earlier. During the period 1983–2002, the arrival of the first few storks was also positively correlated with local temperatures during February and March. Some White Storks have also begun to overwinter in southern Europe, rather than migrating to sub-Saharan Africa. These birds thus have a shorter distance to travel north in spring and may be more responsive to specific weather conditions in areas close to their breeding grounds. Milder winter conditions associated with climatic change may contribute to this increasing tendency of storks to overwinter. Farther east, however, White Storks first appeared in spring in Lithuania about 6 days later during the period 1990–2000 than during the period 1971–1987. On Jurmo Island and the Hanko Peninsula of southern Finland, the Great Crested Grebe also showed a trend for earlier passage during positive phases of the NAO.

Some waterbirds are nesting earlier, as well. In Estonia, the Common Crane advanced its nesting schedule by about 12 days over the twentieth century. The earlier nesting has probably resulted from climatic warming and has contributed to the increase in the population of this species throughout Europe in recent decades.

## Asia

In eastern Asia, projections of habitat change due to climatic warming indicate that the Siberian Crane, which breeds in tundra of Arctic Russia and Siberia and winters in China, will likely lose 70 percent of its breeding habitat. Drought may also affect its wintering habitat in China. Increasing drought is thought to be one factor in the disappearance of Siberian cranes from Keoladeo National Park, India.

## Australia

With climatic warming during the period 1973–2000, several waterfowl and other waterbirds have modified their migration schedules in south-western Australia. Over these years, mean annual temperature in the region increased about 0.31°C per decade. Near Manjimup, various waterbirds showed diverse changes in migration schedules. The Musk Duck and Australasian Grebe extended their summer stays, largely by delaying fall departures. The White-eyed Duck, a diving duck, arrived and departed earlier. The Black Swan, on the other hand, arrived later and departed later. The Great Egret, a wintering species, extended its winter stay, largely by a later departure. The Straw-necked Ibis, a nonbreeding visitor, tended to extend its summer presence by an earlier arrival. As might be expected, the arrival and departure dates, as well as duration of stay, of most waterbirds were significantly related to specific temperature and rainfall measures, although they varied from species to species.

## Southern Africa

The Greater Flamingo, a nomadic species that is highly dependent on rainfall conditions, breeds irregularly at Etosha Pan, Namibia. Successful breeding is dependent on heavy rains that create long-lasting waters favorable for feeding. In only 14 of the years between 1956 and 2006 were birds able to fledge significant numbers of young. Over this period, rainfall declined by about 11 percent. Climatic conditions are predicted to become drier and warmer through this century, leaving the future of flamingo breeding at Etosha and other nearby pans in jeopardy. Changes in the amount and variability of rainfall in the Karoo region of northeastern Cape Province, South Africa, are also likely to influence the nomadic regional population of Blue Cranes.

## Summary

Populations of most ducks, geese, and swans in North America are now at high levels. Warming and drying of the north-central Great Plains of North

America induced by climate change, however, will greatly reduce breeding habitat for many species of dabbling and diving ducks. Sea level rise will also reduce breeding habitat in Arctic coastal areas and also wintering habitat along the Atlantic and Gulf of Mexico coasts. Many species of waterfowl and other waterbirds are now showing earlier spring arrival in breeding areas in North America. The warm phase of the PDO has favored some waterbirds, such as the Whooping Crane. In the Old World, climate models suggest that populations of several geese will decline substantially. Some European waterfowl and storks have advanced their spring arrivals on breeding grounds; others have not. East Asian species such as the Siberian Crane also face challenges from loss of breeding and wintering habitat. Several waterbirds of southwestern Australia have altered their migration schedules.

## KEY REFERENCES

Abraham, K. F., R. L. Jeffries, and R. T. Alisaukas. 2005. "The dynamics of landscape change and snow geese in mid-continent North America." *Global Change Biology* 11:841–855.

Bauer, S., M. Van Dinther, K.-A. Högda, M. Klassen, and J. Madsen. 2008. "The consequences of climate-driven stop-over sites changes on migration schedules and fitness of Arctic geese." *Journal of Animal Ecology* 77:654–660.

Bethke, R. W. and T. D. Nudds. 1995. "Effects of climate change and land use on duck abundance in Canadian prairie-parklands." *Ecological Applications* 5(3):588–600.

Bradley, N. L., A. C. Leopold, J. Ross, and W. Huffaker. 1999. "Phenological changes reflect climate change in Wisconsin." *Proceedings of the National Academy of Sciences USA* 96:9701–9704.

Dickey, M.-H., G. Gautier, and M.-C. Cadieux. 2008. "Climatic effects on the breeding phenology and reproductive success of an arctic-nesting goose species." *Global Change Biology* 14:1973–1985.

Glick, P. 2005. *The Waterfowler's Guide to Global Warming*. National Wildlife Federation, Washington, DC. Warming.pdf.

Jensen, R. A., J. Madsen, M. O'Connell, M. S. Wisz, H. Tømmervik, and F. Mehlums. 2008. "Prediction of the distribution of Arctic-nesting pink-footed geese under a warmer climate scenario." *Global Change Biology* 14:1–10.

Johnson, W. C., B. V. Millett, T. Gilmanov, R. A. Voldseth, G. R. Gunterspergen, and D. E. Naugle. 2005. "Vulnerability of northern prairie wetlands to climate change." *BioScience* 55(10):863–872.

Ptaszyk, J., J. Kosicki, T. H. Sparks, and P. Tryjanowski. 2003. "Changes in the timing and pattern of arrival of the White Stork (*Ciconia ciconia*) in western Poland." *Journal of Ornithology* 144:323–329.

Sanderson, F. J., P. F. Donald, D. J. Pain, I. J. Burfield, and F. P. J. von Bommel. 2006. "Long-term population declines in Afro-Palearctic migrant birds." *Biological Conservation* 131:93–105.

Sorenson, L. G., R. Goldberg, T. L. Root, and M. G. Anderson. 1998. "Potential effects of global warming on waterfowl populations breeding in the northern Great Plains." *Climatic Change* 40:343–369.

Zöckler, C. and I. Lysenko. 2000. Waterbirds on the edge: Climate change impact on Arctic breeding waterbirds. Pp. 20–25 in R. F. Green, M. Harley, M. Spaulding, and C. Zöckler (Eds.), *Impacts of Climate Change on Wildlife*. World Conservation Monitoring Centre Biodiversity Series, No. 11, Cambridge, UK.

# Chapter 14

# *Oceanic Birds: Northern Atlantic, Baltic, and Mediterranean Regions*

Changing climate is modifying atmospheric and oceanic circulation patterns throughout the Northern Hemisphere. Superimposed on these trends are decadal cycles of climate, such as the North Atlantic Oscillation (NAO), that are not easily understood or predicted. Migratory seabird populations in the northeastern Atlantic are showing major changes in response to these patterns, but significant responses to changes in climate and oceanic currents also are being seen in the northwestern Atlantic and elsewhere. Human harvests of marine fish and invertebrates further complicate our efforts to understand the major changes that are occurring in populations of migratory seabirds.

One recent study examined time series of adult survival and reproductive success for seabirds throughout the North Atlantic region in relation to parameters of the NAO. These time series covered periods from the 1960s through the early 2000s, species belonging to six families of seabirds, and an area from South Carolina to New Brunswick, Canada, in the western Atlantic and from Madeira to northern Norway in the eastern Atlantic. Although this broad analysis failed to detect general relationships to the NAO for all of these bird taxa, as a group, strong evidence exists for substantial effects on many seabirds, particularly pelagic species, in several more-limited parts of the North Atlantic region.

## Northwestern Atlantic Region

In the northwestern Atlantic Ocean, the ecology of coastal waters and the diets of seabirds are controlled by the interplay of water currents differing in temperature. This interplay has led to a changing pattern of fish harvest by Northern Gannets and other marine birds. When sea surface temperatures increased by about 0.6°C off the coast of Newfoundland, Canada, over the period 1910–1950, for example, Northern Gannets increased greatly in numbers because of the influx of mackerel, a warm-water fish, into local waters.

In the early 1990s, however, strong cold-water incursions led to a shift from mackerel back to cold-water fish, primarily Capelin (*Mallotus villosus*) in the gannet diet. Capelin, a small cold-water fish of the smelt family (Osmeridae), is one of the most important prey fish for seabirds in the northwest Atlantic. Feeding schools of Capelin are concentrated in the cold margins of Arctic water masses. These fish are very sensitive to temperature changes, with warming of about 1.0°C resulting in a distribution shift of hundreds of kilometers toward colder waters. During most of the 1990s, weakening of the Labrador Current, associated with stronger and more-frequent positive phases of the NAO, generally pushed the distribution of marine plankton and Capelin northward along the coasts of Labrador and Newfoundland.

Weakening of the Labrador Current thus appears to be associated with reduced availability of Capelin relative to other fish utilized by seabirds. On the Gannet Islands off Labrador, Atlantic Puffins fed their young 50–70 percent less Capelin in 1996–1998 than in 1981–1983. Alternative prey, however, apparently sustained reproduction by this puffin population. Similarly, Common Murres and Thick-billed Murres in the Gannet Islands switched from feeding young about 80 percent Capelin in 1981–1983 to less than 50 percent in 1996–2005. For Thick-billed Murres, less than 10 percent of the diet consisted of Capelin in 1996–2005. Nevertheless, sharp changes in prey availability and use can occur from year to year. At Funk Island, off Newfoundland, for example, puffins fed their young with sand lance (*Ammodytes dubius*) 94 percent of the time in 2004 and large Capelin 83 percent of the time in 2005.

During the 1990s, seabirds off the coasts of Labrador and Newfoundland also tended to delay their nesting. In the Gannet Islands, for example, egg laying by Black-legged Kittiwakes and murres was later during 1996–1998 than in the early 1980s. In some bird colonies, late appearance of Capelin led to reduced survival of chicks hatching early in the season. In spite of the changes in diets of the various seabirds, however, overall populations have apparently not declined.

Winter diets of murres in this region have also changed. Arctic Cod (*Boreogadus saida*) declined from 55 to 12 percent and Capelin from 28 to 6 percent between 1984–1986 and 1996–1998 in waters off Newfoundland. In addition, invertebrate foods such as euphausid shrimp, present in over 48 percent of stomachs examined in 1984–1986, were completely absent in 1996–1998.

Other seabirds appear now to be extending their breeding range northward onto the coast of Labrador. These include the Caspian Tern, Double-crested Cormorant, and Black-headed Gull.

In the Canadian high Arctic Ocean, the population of the Ivory Gull has declined about 80 percent since the 1980s. Because this species is a year-round associate of pack ice, the influence of warming climate on ice conditions that are essential to successful foraging may well be involved in the population decline. This gull also scavenges food from Polar Bear kills, so changes in bear activity and reduced availability of such food may be a contributing factor. Changes in sea ice conditions in their presumed wintering grounds between Greenland and Canada may also be reducing their feeding success.

In Greenland, seabird populations have crashed in some places, but the role of climate change is uncertain. Counts of Black-legged Kittiwakes from 1920 to 1999 indicate that the species has declined in numbers at thirty-four of fifty major colonies, while increasing at only five. The cause of these declines was not known. Seabird surveys in 2000 on the west coast of Greenland between 70° and 72° N revealed that nesting colonies of Thick-billed Murres, Black-legged Kittiwakes, and Razorbills, as well as populations of Common Eiders, had decreased by more than 50 percent. Thick-billed Murres, estimated at 500,000 pairs a century earlier, had completely disappeared. Hunting and harvest of eggs by humans were considered the likely causes of these declines. About 55,000–70,000 eiders are apparently killed annually by hunting or bycatch in gill nets in Greenland.

## Hudson Bay

Hudson Bay has experienced gradually warming weather and reduced extent of summer ice since about 1986. Correlated with these changes, the colony of Thick-billed Murres at Coats Island, in the northern part of the bay, has advanced its median egg-laying date. The median date is also strongly correlated with temperatures during May and June; the warmer the temperature, the earlier the laying date. During this period, the fish fed to the young murres has shifted from largely Arctic Cod, a fish characteristic of Arctic waters, to Capelin and sand lance, species typical of low Arctic waters. In the early 1980s, cod made up about 51 percent of foods for murre chicks, but it was only 19 percent by 1999. In years with the lowest ice cover near the colony, body mass of murre chicks was also substantially reduced.

With current projections of continued warming in the Hudson Bay area, it appears likely that Thick-billed Murres will shift their principal breeding sites northward. At the same time that the Coats Island birds were experiencing the changes noted above, those at Prince Leopold Island, in the high Arctic, had shown little directional change in breeding time or chick growth.

Although breeding has not yet occurred, Razorbills have begun to visit the Coats Island seabird colonies, with twelve individuals recorded in 2002. This site is 300 kilometers west of the nearest previously known breeding colony. This species also forages heavily on sand lance.

## Northeastern Atlantic Region

In the northeastern Atlantic, warming of ocean waters, coupled with the tendency of the NAO to maintain a positive phase that promotes warm, wet, stormy winters, has led to major changes in marine ecology. From Iceland eastward, but particularly in the North Sea between the British Isles and mainland Europe, many migratory seabird populations have declined. The species most severely affected include the Northern Fulmar, Black-legged Kittiwake, Parasitic Jaeger, Thick-billed Murre, Common Murre, Atlantic Puffin, Arctic Tern, Little Tern, and European Shag.

Responses of these seabird populations almost certainly reflect basic patterns of change in food resources. In this oceanic region, cold-water plankton has shifted north by about 1000 kilometers since the 1930s. This change, together with commercial fishing, has led to reductions in breeding success and survival of migratory seabirds.

Since the 1980s, populations of Black-legged Kittiwakes, Thick-billed Murres, European Shags, Northern Fulmars, and other seabirds have declined throughout much of the North Sea. Recently, the Shetland Islands have seen the most serious seabird declines in the United Kingdom. These declines have accelerated since the mid-1980s. By 1998–2002, Black-legged Kittiwake numbers had fallen by 62 percent, Parasitic Jaeger by 42 percent, and Arctic Tern by 19 percent. In the Orkney Islands, just north of Scotland, Northern Fulmars have declined in numbers since 2000, after a long period of population growth.

In 2004, almost complete breeding failures of seabirds were noted in the Shetland Islands, the Isle of May east of Scotland, and the northeastern British coast. Black-legged Kittiwake, Parasitic Jaeger, and Arctic Tern populations in the Shetland Islands fell to numbers lower than those noted in 2002. These failures were prefaced in February 2004 by appearances of Northern Fulmars in inland areas of Britain, and by hundreds of dead fulmars washing up on coastlines of England, France, Belgium, Germany, and the Netherlands. These birds had apparently died of starvation.

In 2005, some recovery of breeding seabirds occurred in the areas worst affected in 2004, but reduced breeding success spread to colonies in northwestern Scotland. In 2006, seabird colonies along the western coast of Britain showed very low reproductive success. Late in 2007, heavy mortality on many seabirds, particularly Razorbills, was reported along the coasts of Norway, Sweden, and Denmark. Many were extremely emaciated and appeared to have died of hunger.

The proximate cause of these failures appears to be the collapse of Lesser Sandeel (*Ammodytes marinus*) populations, a plankton feeder in waters from the Shetland Islands east across the North Sea. The decline of Black-legged Kittiwakes in Shetland Island and North Sea breeding colonies during the 1990s is linked closely to changes in availability of this fish. By the late 1990s, the large kittiwake colony at the Isle of May was declining by about 11 percent annually. During this period, a major commercial fishery for sandeels operated in the North Sea, harvesting up to a million tons annually. The operation of this fishery has been restricted since 2000. But in addition to the fishery, warmer sea waters, especially during winter when sandeels spawn, have reduced recruitment by sandeels. Modeling of population dynamics has suggested that sea temperature has played a significant role in the continued relatively low breeding success and survival of kittiwakes. Modeling also indicated that continued increase in winter water temperatures is likely to further reduce sandeel recruitment and kittiwake populations.

The Barents and Norwegian Sea coasts of Norway are major seabird breeding areas, where about 1.3 million pairs of seabirds of eighteen species nest. Here, as well, populations of the Common and Thick-billed Murres,

TABLE 14.1. Changes in populations of seabirds on the northern Barents Sea coast and central Norwegian Sea coast of Norway prior to 1995 and between 1996 and 2005.

|  | Annual Percent Change | |
|  | Barents Sea | Norwegian Sea |
| --- | --- | --- |
| Black-legged Kittiwake | | |
| < 1995 | −2.2 | −3.3 |
| 1996–2005 | −6.4 | −7.8 |
| Common Murre | | |
| < 1995 | −14.8 | −5.7 |
| 1996–2005 | 10.3 | −19.7 |
| Thick-billed Murre | | |
| < 1995 | −14.3 | — |
| 1996–2005 | −25.9 | — |
| Atlantic Puffin | | |
| < 1995 | 2.4 | −1.9 |
| 1996–2005 | 1.9 | −2.1 |
| Razorbill | | |
| < 1995 | — | — |
| 1996–2005 | −2.9 | −8.4 |
| Lesser Black-backed Gull | | |
| < 1995 | — | −5.4 |
| 1996–2005 | — | −3.2 |

*Source:* Data from Barrett et al. 2006.

Atlantic Puffin, Razorbill, Black-legged Kittiwake, and Lesser Black-backed Gull are declining rapidly in numbers (Table 14.1). The breeding season of puffins is also influenced by the NAO, beginning earlier in years when the NAO is in a positive phase.

The timing of reproduction of many of these seabirds has changed. Black-legged Kittiwakes and Thick-billed Murres at the Isle of May, south-eastern Scotland, showed a trend of later breeding over the period from the early 1980s through 2002. Both of these species rely heavily on the Lesser Sandeel during the breeding season. The European Shag, although also a sandeel feeder, showed little change in breeding period from the late 1960s through 2002.

A number of studies have also suggested that mild winter temperatures have contributed to seabird population declines due to low survival at sea. The species involved include Atlantic Puffins, Common Murres, and

Northern Fulmars. Warm ocean conditions are thought to reduce recruitment of fish stocks on which these birds depend.

Another disturbing development, beginning in 2003, is the spread of the Snake Pipefish (*Entelurus aequoreus*) into the North Sea. The species has suddenly emerged as one of the most explosive of the southern fishes that have invaded the warming waters of the North Sea. This heavily armored fish, with a stiff vertebral column, is apparently easy for seabirds to catch, but it is difficult to swallow, especially for young birds. Some young birds have starved to death while sitting on uneaten pipefish. Young Arctic Terns have also choked to death while trying to swallow pipefish. Pipefish are also lower in food energy value and more difficult to digest than other fish, particularly sandeels.

Ranges of several gulls have changed in apparent correlation with climatic warming. The Lesser Black-backed Gull populations in Britain and Ireland have increased, and the species has become a common wintering bird. Much of this increase is correlated with decrease in abundance of the Herring Gull. Wintering Lesser Black-backed Gulls have increased exponentially in Ireland since the late 1970s. Increasingly mild winters may have contributed to this increase.

The Mediterranean Gull has expanded its breeding range from the Mediterranean and Black seas to the European coast and southern England. In the late 1980s only a handful of breeding attempts were made in England, but the number of breeding birds has risen sharply, and in 1999–2002 some 108 pairs nested in Britain and 5 pairs in Ireland.

## Baltic Sea

In the Baltic Sea region, climatic warming is projected to cause the breeding and wintering ranges of seabirds generally to shift northward. Some species are also advancing their breeding schedules. Over the period 1929–1998, Arctic Terns advanced their breeding by about 18 days at colonies in Denmark. This change is positively correlated with an increase in mean temperature during April and May and positively correlated with the NAO Index for May. Over this period, natural selection appeared to favor early-breeding terns, whose young showed greater survivorship. In addition, the dispersal pattern of young birds changed during the 70-year period. In the 1930s, the first breeding sites of terns averaged about 10 kilometers from their natal colony, but by the 1990s they had increased to an average of 100 kilometers.

## Mediterranean Sea

Foraging areas of nonbreeding pelagic seabirds are also changing, although data to document such changes are scarce. The Balearic Shearwater nests in the Balearic Islands of the western Mediterranean. This shearwater is listed as critically endangered in the 2007 Red Data List of the International Union for Conservation of Nature, its population having declined to about 2000 pairs. Recently, however, it appears to have expanded its nonbreeding range in the northeastern Atlantic, correlated with warming of the waters there. Based on 25 years of data, a report from one group of observers told of a rapid northward expansion of the nonbreeding range of this plankton and fish feeder. Increasing temperature of surface waters and correlated northward shifts of plankton and prey fish in the mid-1990s are interpreted as the controlling factors. The range of this species has thus expanded northward by about 644 kilometers. The birds are now commonly seen in waters off the southern British Isles, where they were rare only 20 years ago. Although the reality of this expansion is questioned by some, recent evidence suggests that these nonbreeding birds are ranging as far north as Scandinavia, Scotland, and Wales.

## Summary

In the northwestern Atlantic Ocean, weakening of the Labrador Current has led to warming of coastal waters and a northward shift of the fish prey of seabirds, particularly Capelin. Seabirds have altered their diets but maintained normal populations, although nesting has become later in some cases. In Hudson Bay, climatic warming has enabled some seabirds to nest earlier in the season, and northward shifts of breeding areas by some species are expected. In the northeastern Atlantic, however, warming of waters from Iceland to the North Sea has led to major declines of many seabirds. In the Baltic Sea, Arctic Terns have begun nesting earlier in the season. The Balearic Shearwater has expanded its foraging range from the Mediterranean Sea to the eastern Atlantic.

### KEY REFERENCES

Barrett, R. T., S.-H. Lorentsen, and T. Anker-Nilssen. 2006. "The status of breeding seabirds in mainland Norway." *Atlantic Seabirds* 8(3):97–126.

Carscadden, J. E., W. A. Montevecchi, G. K. Davoren, and B. Nakashima. 2002. "Trophic relationships among capelin (*Mallotus villosus*) and seabirds in a changing ecosystem." *ICES Journal of Marine Sciences* 59:1027–1033.

Davoren, G. K. and W. A. Montevecchi. 2003. "Signals from seabirds indicate changing biology of capelin biology." *Marine Ecology Progress Series* 258:253–261.

Durant, J. M., T. Anker-Nilssen, D. Ø. Hjermann, and N. C. Stenseth. 2004. "Regime shifts in the breeding of an Atlantic Puffin population." *Ecology Letters* 7:388–394.

Frederiksen, M., S. Wanless, M. P. Harris, P. Rothery, and L. J. Wilson. 2004. "The role of industrial fisheries and oceanographic change in the decline of North Sea black-legged kittiwakes." *Journal of Animal Ecology* 41:1129–1139.

Gaston, A. J., H. G. Gilchrist, and J. M. Hipfner. 2005. "Climate change, ice conditions and reproduction in an Arctic nesting marine bird: Brunnich's Guillemot (*Uria lomvia* L.)." *Journal of Animal Ecology* 74:832–841.

Harris, M. P., T. Anker-Nilssen, R. H. McCleery, K. E. Erikstad, D. N. Shaw, and V. Grosbois. 2005. "Effect of wintering area and climate on the survival of adult Atlantic puffins *Fratercula arctica* in the eastern Atlantic." *Marine Ecology Progress Series* 297:283–296.

Harris, M. P., D. Beare, R. Toresen, L. Nöttestad, M. Kloppmann, H. Dörner, K. Peach, D. R. A. Rushton, J. Foster-Smith, and S. Wanless. 2007. "A major increase in snake pipefish (*Entelurus aequoreus*) in northern European seas since 2003: Potential implications of seabird breeding success." *Marine Biology* 151:973–983.

Moline, M. A., N. J. Karnovsky, Z. Brown, G. J. Divoky, T. K. Frazer, C. A. Jacoby, J. J. Torres, and W. R. Fraser. 2008. "High latitude changes in ice dynamics and their impact on polar marine ecosystems." *Annals of the New York Academy of Sciences* 1134:267–319.

Möller, A. P., E. Flensted-Jensen, and W. Mardal. 2006. "Rapidly advancing laying date in a seabird and the changing advantage of early reproduction." *Journal of Animal Ecology* 75:657–665.

Wanless, S., M. Frederiksen, F. Daunt, B. E. Scott, and M. P. Harris. 2007. "Black-legged kittiwakes as indicators of environmental change in the North Sea: Evidence from long-term studies." *Progress in Oceanography* 71(1):30–38.

Wynn, R. B., S. A. Josey, A. P. Martin, D. G. Johns, and P. Yésou. 2007. "Climate-driven range expansion of a critically endangered top predator in northeast Atlantic waters." *Biology Letters* 3:529–532.

# Chapter 15

# *Oceanic Birds: North Pacific*

Major changes have occurred in the marine ecosystems in several regions of the northern Pacific. These changes involve long-term oscillations of climate and shifts in ocean currents that influence primary productivity, fish populations, and the populations of many marine birds. Ocean warming and a decrease in the extent of sea ice in the Bering Sea and Arctic Ocean are influencing marine food webs leading to many seabirds.

Influences of the El Niño Southern Oscillation also extend into both eastern and western parts of the North Pacific. These have led to reproductive failures by seabirds along the California coast and around Japan. The Pacific Decadal Oscillation also influences marine food chains leading to seabirds throughout most of the North Pacific. Changing global climate is altering the influence of at least the El Niño Southern Oscillation.

Climatic warming is also leading to rapid rise in sea level, which has major implications for many seabirds breeding on low-lying ocean islands in the tropical Pacific.

## Arctic Ocean and Bering Sea

The seabird colonies of the Bering and Chukchi seas are among the most spectacular on earth. At Cape Thompson, Alaska, north of Kotzebue, where I worked as a member of a seabird study group in 1960 and 1961, the cliffs were the nesting site of over 350,000 murres, puffins, gulls, cormorants, and guillemots. These cliffs bordered the site chosen by the Atomic Energy Commission for Project Chariot, which proposed the use of a series of five nuclear explosions to create an inlet and harbor. How these blasts would have affected the nesting cliffs is uncertain, but our seabird group was happy that the project was eventually cancelled.[1]

In the Bering Sea and Arctic Ocean, some of these seabirds are now being influenced by progressive climatic change that affects coastal nesting areas and oceanic foraging areas associated with the ice pack. The best-documented case involves the Black Guillemot, a circumpolar species that nests in cavities of coastal cliffs, including those at Cape Thompson. The nesting cycle, from parental occupation of the nest cavity to fledging of the chick, lasts about 80 days. If early autumn snows begin before this period is completed, the chicks may be trapped in their nest cavities and die. This limits the distribution of the species to coastal areas with a snow-free period at least this long. In areas long utilized for nesting by this species, the date of egg laying has advanced about 4.5 days per decade as the climate has warmed. The snow-free period at Point Barrow has exceeded 80 days on a regular basis since about the mid-1960s.

As a result, breeding areas of Black Guillemots have expanded northward to localities such as Cooper Island, near Point Barrow, where breeding was unknown prior to the 1960s. In 1972, a nesting colony of 10 pairs of Black Guillemots was discovered on Cooper Island. Between 1975 and 1990, this colony grew to 225 pairs. Timing of egg laying has also advanced. From 1975 through 2002, egg laying became earlier at a rate of about 3 days per decade (Figure 15.1).

On the other hand, since Arctic Black Guillemots are obligate foragers at the edge of the ice pack throughout the year, feeding primarily on Arctic Cod (*Boreogadus saida*), their breeding success is related to the location of sea ice in summer. Reproductive success is inversely related to the distance of the ice pack from coastal nesting areas in the later part of the breeding cycle. The numbers of guillemots at Cooper Island increased steadily during the 1970s and 1980s but declined substantially in the 1990s, correlated with the general recession of Arctic sea ice and increased distance between nesting and foraging areas. By 1997, the colony of Black Guillemots had declined from 225 to 110 pairs, apparently because of reduced immigration and increased mortality. Chukchi Sea populations of Kittlitz's Murrelet also appear to be declining as a result of retreat of the pack-ice front.

---

[1] See Dan O'Neill. 1994. *The Firecracker Boys*. St. Martin's Press, New York.

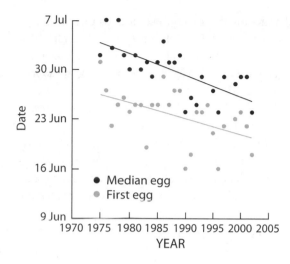

FIGURE 15.1. First and median egg-laying dates for Black Guillemots at Cooper Island, Alaska, from 1975 through 2002. (Figure modified from www.cooperisland .org/importantfindings.htm by permission of the Friends of Cooper Island.)

Horned Puffins and Tufted Puffins are also nesting farther north along the northwest Alaska coast. The Horned Puffin also requires a snow-free period of at least 90 days, which, as we noted, now occurs frequently in the Point Barrow area. The first breeding of this species occurred at Cooper Island in 1986. Although warming conditions have permitted northward expansion of breeding localities for these puffins, continued retreat of the sea ice may be detrimental. As the edge of the ice pack becomes more distant from breeding colonies in summer, these birds, like Black Guillemots, may find decreased food availability within normal foraging distance.

In the Bering Sea and parts of the western North Pacific, including coastal bay waters of the Sea of Okhotsk, shifts of water temperature occurred in the latter part of the 1900s. These led to changes in the abundance of zooplankton and fish that are reflected in changes in reproductive success of planktivorous auklets and piscivorous puffins and murres. Warmer water inflows favored smaller forms of plankton, the prey of juvenile fish, and thus were beneficial to the piscivorous Horned and Tufted Puffins. Cold waters favored larger plankton on which the Crested and Parakeet Auklets fed. These conditions prevailed in the Bering Sea between the mid-1950s and the late 1970s. Warmer ocean waters since the late 1970s, apparently due to a shift in the Pacific Decadal Oscillation, appear to be benefiting piscivorous seabirds.

Major changes are occurring in sea ice conditions and evidently in marine food chains in the Pribilof Islands of the eastern Bering Sea. Since 1997, winter sea ice has occurred less frequently in the area of the Pribilofs, and the changed pattern of sea ice retreat in spring influences the manner in which algae associated with the ice margin feed into benthic and pelagic food chains. Between 1975 and 2006, both Black-legged and Red-legged Kittiwakes have advanced their nesting dates substantially. On St. Paul Island, on the other hand, Thick-billed Murres have delayed nesting by about half a day per year. In addition, the number of chicks fledged per pair declined by about 50 percent. These changes apparently reflect changes in pelagic food webs leading to kittiwakes and in benthic food webs leading to murres.

Changes in ocean temperature are also affecting Least Auklets in the Pribilof Islands. From 2002 to 2005, warm ocean conditions led to a severe shortage of copepods on which this species depends. Hormone levels of birds from Pribilof colonies showed that the birds were under nutritional stress, and they possibly suffered poor reproductive success. Continued warming trends in the eastern Bering Sea are likely to reduce populations of Least Auklets and other planktivorous seabirds

## Gulf of Alaska

In the Gulf of Alaska, the shift from cold to warm surface water that began in 1977 has altered the food supply of seabirds. Capelin (*Mallotus villosus*) was the principal prey fish for seabirds prior to this time. Shrimp (*Pandalus* spp.) and Capelin have since declined to very low levels. Decline of Capelin, in particular, was associated with shifts in the diets of several marine birds and mammals that fed heavily on them and, ultimately, with the decline of many bird and mammal populations, particularly of Common Murre. Breeding success of Black-legged Kittiwakes declined heavily throughout the Gulf of Alaska during the 1980s. In the Prince William Sound area, populations of many seabirds declined between 1972–1973 and 1989–1991, in areas both affected and unaffected by the Exxon *Valdez* oil spill. These included Pigeon Guillemots, Marbled and Kittlitz's Murrelets, and Horned Puffins and a number of gulls. Episodes of heavy mortality have occurred, including starvation deaths of an estimated 100,000 Common Murres in March 1993.

Farther west, in the west-central Aleutian Island region, Least Auklets have also shown survival patterns related to long-term changes in oceanic productivity. Auklets in colonies at Buldir Island during the period 1990–2000 have shown declining survivorship, correlated with increasing values of the North Pacific Index, which in turn reflects a gradually increasing atmospheric pressure, warming sea-surface temperatures, and declining ocean primary productivity. During the nonbreeding season, Least Auklets disperse away from nesting islands and are apparently influenced strongly by food availability in the open ocean. At the same locality, however, Whiskered Auklets, which remain near shore during the nonbreeding season, experienced less variation in food availability but were more seriously affected by winter storms. Their survival was lowest during winters with weather dominated by strong low atmospheric pressure.

## Canadian and Western United States Waters

Farther south, along the coast of British Columbia, the survival and reproduction of several species of seabirds have suffered because of changes in food availability. These changes are tied to the general increase in water temperatures that began in the 1970s. At Triangle Island, off the northern tip of Vancouver Island, along the coast of British Columbia, Canada, Cassin's Auklets and Rhinoceros Auklets experienced low survival rates during the period 1994–1997. The peak of zooplankton abundance tended to occur prior to Cassin's Auklet breeding, leading to poor chick growth and reduced reproductive success. Survival of Cassin's Auklets at the Triangle Island colony was lower than that at Reef Island and Frederick Island, farther north along the British Columbia coast and north of the area under the influence of the California Current. The low survival rates suggested that decreased food availability in coastal waters was likely.

Studies of reproductive success and diet of Rhinoceros Auklets at Triangle Island showed that availability of sand lance (*Ammodytes hexapterus*), a small fish abundant in colder waters but unavailable when warm waters occupy the coastal area, was critical to successful breeding. Although a number of other small fish were also fed to nestlings, sand lance constituted about 38 percent, on average, and in some years up to 86 percent.

In 2005, colonies of Cassin's Auklets at both Triangle Island, British Columbia, and the Farallon Islands, California, suffered nearly complete reproductive failures. These failures apparently resulted from the absence of near-coastal upwelling due to atmospheric wind conditions that permitted warm surface waters to dominate the coastal region. Breeding by Cassin's Auklets failed again at the Farallon Islands in 2006. The birds returned and nested in 2007, with about a third of the pairs fledging young. Adult survival and recruitment of first-time breeders has also declined in this colony.

Tufted Puffins have also suffered from warming of the ocean waters in which they forage during the breeding season. Between 1975 and 2002, the largest colony of these birds along the British Columbia coast, located on Triangle Island, experienced changing phenology and increasingly variable reproductive success. Sea surface temperatures since the 1940s show a general warming trend, with considerable variability and decadal swings toward colder waters in the 1950s and 1970s. With warming conditions, breeding has become earlier, with the mean hatching date for chicks advancing from July 16 in 1975–1982 to June 30 in 1994–2002. Nestling growth rates are strongly affected by water temperatures and are lowest when waters are warmest; in years when warm water conditions prevail throughout the nesting season, fledging success is low or almost zero. Normal growth and successful fledging depend on the availability of sand lance. Tufted Puffin colonies in Oregon and Washington have also declined severely.

Populations of Marbled Murrelets, a partial migrant, have apparently declined by about 71 percent along the coasts of Alaska and British Columbia. Although several factors have contributed to population decline, including mortality in fishing nets, oil spills, and old-growth forest harvest, climate-induced changes in the marine ecosystem since the late 1970s have likely played a role. This marine shift reduced the abundance of small forage fish preferred by murrelets, replacing them by larger predatory fish. Studies using stable-isotope analysis also show that murrelets along the entire west coast have switched from feeding on small fish with high caloric value to feeding on krill and other invertebrates with lesser food value, correlated with the marine ecosystem shift. Forest harvest is also detrimental to the species because most birds nest on mossy limbs high in old-growth trees. Fledgling Ancient Murrelets reared in central Hecate Strait, between the Queen Charlotte Islands and mainland British Columbia, showed earlier departures from nesting areas and lighter body masses over the period 1983–1999. These trends were postulated to reflect long-term changes in oceanographic conditions.

The Sooty Shearwater is one of the most abundant and wide-ranging pelagic seabirds in the world. It breeds on islands in the Southern Hemisphere but ranges over most of the Atlantic and Pacific oceans during the nonbreeding period. Birds with electronic tracking tags reveal that in the Pacific, many shearwaters concentrate their feeding in the California Current, the Gulf of Alaska, and ocean areas east and northeast of Japan. In their nonbreeding travels, these birds apparently cover more than 64,000 kilometers annually. With the warming of the surface waters of the California Current between 1949 and 1990, changes in abundance of the Sooty Shearwater also occurred. Nonbreeding shearwaters, which feed largely on euphausiid shrimp, larval fish, and squid, declined in numbers between 1976 and 1998. A moderate increase has occurred in the California Current since 1999. Since this is a major portion of the nonbreeding range of this species, and increases in their numbers do not seem to have occurred elsewhere, the total population of the species still appears to be depressed.

## Tropical North Pacific

In tropical regions, many seabirds nest on low-lying coral atolls, which are at risk from rising sea levels (see Chapter 3). The Northwestern Hawaiian Islands, for example, are home to about 14 million seabirds of eighteen species. The total land area of these islands is only about 800 hectares, and a sea level rise of about half a meter would likely reduce their area by more than 50 percent by AD 2100. Some losses of nesting islets have already occurred. Whale Skate Island, one of the small islands that form the French Frigate Shoals, was about 10 to 15 acres in size. It was vegetated and supported colonies of nesting seabirds and was also used by Hawaiian monk seals and nesting green sea turtles. Erosion, apparently due to rising sea level, caused the island to disappear over the course of 20 years. Beaches throughout the Hawaiian Islands are suffering from increased erosion.

In the tropical Pacific, the abundance of planktivorous and piscivorous seabirds is also influenced by thermal boundaries between the South Equatorial Current and the North Equatorial Countercurrent. Changes in current patterns related to the El Niño Southern Oscillation and general warming of tropical oceans may thus influence the relative abundances of tropical seabirds, particularly plankton-feeding birds such as petrels, storm-petrels, and phalaropes.

## Summary

Warming of the Bering Sea and Arctic Ocean, combined with northward recession of the ice pack, have permitted the northward expansion of nesting areas for some seabirds. The breeding success of birds that feed in close association with the ice pack, on the other hand, has decreased in some locations. Increase in water temperatures along the western coasts of British Columbia and the United States have led to reproductive failures of a number of seabirds and poor feeding conditions for species such as the Sooty Shearwater. In the Aleutian Islands auklets have experienced differing population trends associated with cyclical weather patterns affecting food availability and survival. Sea level rise may already be affecting low-lying islands used for nesting by seabirds in the tropical Pacific.

KEY REFERENCES

Anderson, P. J. and J. F. Piatt. 1999. "Community reorganization in the Gulf of Alaska following ocean climate regime shift." *Marine Ecology Progress Series* 189:117–123.

Baker, J. D., C. L. Littman, and D. W. Johnston. 2006. "Potential effects of sea level rise on the terrestrial habitats of endangered and endemic megafauna in the Northwestern Hawaiian Islands." *Endangered Species Research* 4:1–10.

Becker, B. H., M. Peery, and S. Beissinger. 2007. "Ocean climate and prey availability affect the reproductive success of an endangered seabird, the marbled murrelet." *Marine Ecology Progress Series* 329:267–279.

Bertram, D. F., A. Harfenist, and B. D. Smith. 2005. "Ocean climate and El Niño impacts on survival of Cassin's Auklets from upwelling and downwelling domains of British Columbia." *Canadian Journal of Fisheries and Aquatic Sciences* 62:2841–2853.

Divoky, G. J. and A. M. Springer. 2006. *Shifting Seabird Distribution and Abundance Reflect a Rapidly Changing Marine Environment in the Western Arctic.* Seabirds as Indicators Symposium, Pacific Seabird Group, USGS.

Ford, R. G., D. G. Ainley, J. L. Casey, C. A. Keiper, L. B. Spear, and L. T. Ballance. 2004. "The biogeographic patterns of seabirds in the central portion of the California Current." *Marine Ornithology* 32:77–96.

Gaston, A. J. and J. L. Smith. 2001. "Changes in oceanographic conditions off northern British Columbia (1983–1999) and the reproduction of a marine bird, the Ancient Murrelet (*Synthliboramphus antiquus*)." *Canadian Journal of Zoology* 79:1735–1742.

Gjerdrum, C., A. M. J. Vallée, C. C. St. Clair, D. F. Bertram, and J. L. Ryder. 2003. "Tufted puffin reproduction reveals ocean climate variability." *Proceedings of the National Academy of Sciences USA* 100:9377–9382.

Jones, I. L., F. M. Hunter, G. J. Robertson, J. C. Williams, and G. V. Byrd. 2007. "Covariation among demographic and climate parameters in Whiskered Auklets *Aethia pygmaea*." *Journal of Avian Biology* 38:450–461.

Moline, M. A., N. J. Karnovsky, Z. Brown, G. J. Divoky, T. K. Frazer, C. A. Jacoby, J. J. Torres, and W. R. Fraser. 2008. "High latitude changes in ice dynamics and their impact on polar marine ecosystems." *Annals of the New York Academy of Sciences* 1134:267–319.

Norris, D. R., P. Arcese, D. Preikshot, D. F. Bertram, and T. K. Kyser. 2007. "Diet reconstruction and historic population dynamics in a threatened seabird." *Journal of Applied Ecology* 44:875–884.

Piatt, J. F., K. J. Burger, S. A. Hatch, V. L. Friesen, T. P. Birt, M. L. Arimitsu, G. S. Drew, A. M. A. Harding, and K. S. Bixler. 2007. Status review of the marbled murrelet (*Brachyramphus marmoratus*) in Alaska and British Columbia. USGS Open-file Report 2006-1387. 258 pp. http://pubs.usgs.gov/of/2006/1387/pdf/ofr20061387.pdf.

Shaffer, S. A., Y. Tremblay, H. Weimerskirch, D. Scott, D. R. Thompson, P. M. Sagar, H. Moller, et al. 2006. "Migratory shearwaters integrate oceanic resources across the Pacific Ocean in an endless summer." *Proceedings of the National Academy of Sciences USA* 103:12799–12802.

# Chapter 16

## *Oceanic Birds: Southern Hemisphere*

Migratory birds of the southern oceans are now experiencing perhaps the most severe impacts of changing global climate. The increasing frequency and severity of the El Niño Southern Oscillation is altering ocean regimes, from equatorial waters to the waters surrounding Antarctica. The Antarctic Peninsula, the subantarctic islands, and the adjacent ocean have experienced the greatest climatic warming yet recorded. The western Antarctic Peninsula has warmed about 6°C, more than any other location on earth. Winter sea ice now forms later in autumn, is less extensive in most years, and recedes earlier.

In subantarctic waters, the changes in populations of some birds are also being influenced by fishing practices such as krill harvest and longline fishing. Recent Antarctic krill harvests by the commercial fishery equal about 100,000 tons annually, but the sustainable harvest that does not endanger krill-feeding vertebrates under the rapidly changing climatic conditions is uncertain. Longline fishing is also a problem. This technique utilizes a main monofilament line, up to 100 kilometers in length, with hundreds or thousands of baited hooks attached along its length. Seabirds tend to be caught on the baited hooks as the longline is released from the fishing boat, and they are dragged underwater and drowned.

## Productivity of Foods Required by Seabirds

Major declines have occurred in the abundance of krill, which is a crucial link in the food chains leading to almost all marine vertebrates. About 50 percent of krill in Antarctic waters was previously concentrated in the southwest Atlantic Ocean sector between the tip of South America, South Georgia, and the Antarctic Peninsula. Since the 1970s, krill has declined substantially in this area, as well as in the waters bordering the Antarctic Peninsula, while salps—pelagic gelatinous tunicates—have increased in abundance. As we noted in Chapter 5, salps have very low food value for marine vertebrates.

Decline in the area of sea ice is directly correlated with the decline in krill abundance. The undersurface of the thin sea ice is the site for growth of ice algae that are nourished by upwelling of nutrient-rich water at the edges of the continental shelf. These algae are food for larval krill, so alteration of the extent and seasonal presence of sea ice has a major influence on krill production.

## The Antarctic Peninsula and Adjacent Islands

Along the Antarctic Peninsula, changes in climate and sea ice conditions have affected populations of several species of penguins since the late 1900s. The species that have suffered most are the Adélie and Emperor Penguins, the two most southern Antarctic species. Along the western Antarctic Peninsula, Adélie Penguins, which forage year-round in the pack ice zone, have declined in numbers as the pack ice zone has shrunk. On King George Island, off the northern tip of the Antarctic Peninsula, the Adélie population dropped by half in the 1980s because of reduced chick survival. Many rookeries formerly occupied by Adélies are completely abandoned. Farther south, at about latitude 60°S, the Adélie colonies near Palmer Station on Anvers Island have declined by nearly 70 percent since 1975. In these areas, it appears likely that sea ice extent is well below the optimum for Adélies.

Meanwhile, populations of Chinstrap Penguins, which feed in open water, have increased enormously. Evidence shows that Chinstraps, together with Gentoo Penguins, probably invaded colonies in the northern Antarctic Peninsula between about 1940 and 1970. The prognosis for penguin populations on the Antarctic Peninsula is that Adélie Penguins will continue their southward retreat and Chinstrap and Gentoo Penguins will expand their ranges. Gentoo Penguins were absent from Anvers Island, midway down the peninsula, in the 1950s but have recently appeared at colonies there. Paleoecological evidence indicates that they had not occurred this far south during the past 800 years. Small numbers of Macaroni Penguins, a subantarctic species, have also colonized the Antarctic Peninsula.

The growing populations of Gentoo Penguins and declining numbers of Adélies were apparent when I visited the Antarctic Peninsula in December 2003. Our group went ashore at Petermann Island, on the west side of the peninsula and just across a narrow channel from Anvers Island. There we found about 1000 Gentoo Penguins that were sharing a nesting area with 200 Adélie Penguins and about 14 Antarctic Shags. Brown Skuas were patrolling the colony, hoping to find untended chicks or nests. Although small numbers of Gentoos were present in the Anvers Island area by the 1980s, they were still greatly outnumbered by Adélies, which were estimated to number over 28,000 pairs. The ratio of their numbers now has approximately reversed.

At the Dion Islands, off the southern part of the Antarctic Peninsula, a small colony of Emperor Penguins that was observed for many years has declined from about 150 pairs in the 1950s to 85 pairs in 1978 and to fewer than 10 pairs in recent years. This colony is remarkable in that it is one of the only two Emperor Penguin colonies in which the birds nest on land rather than on ice shelves.

More recently, a colony of Emperor Penguins was discovered at Snow Hill Island, farther north along the west side of the Antarctic Peninsula (latitude 64.3°S). This colony, located on an ice shelf, was estimated to contain 4000–4200 pairs when first visited in 2004. The history of this colony is uncertain, but it may consist of descendants of birds that were reported to be nesting on the Larsen Ice Shelf farther south in the late 1800s. The recent collapse of this ice shelf may have forced birds from this colony to relocate.

## Eastern Antarctica and the Ross Sea Region

Changing sea ice conditions have also affected krill and seabird populations in other locations around Antarctica. The Emperor Penguin, which has about thirty breeding colonies located mainly on ice shelves around the edge of the continent, is one of the species with population dynamics most strongly influenced by sea ice conditions. At the Dumont d'Urville colony

in Terre Adélie, east of the Ross Sea, Emperor Penguin populations declined by about 50 percent during the late 1900s, mostly because of high
adult mortality in the late 1970s. This adult mortality was correlated with
reduced sea ice cover, which, in turn, was related to reduced krill abundance
and recruitment in areas where adult penguins forage. Adult mortality may
also be related to unstable pack-ice conditions during the summer molting
period, when birds must remain out of the water on sea ice for 4 weeks during the summer. Breeding failure also occurs in years when breakup of ice
on which the colony is located occurs, or when severe blizzards strike the
colony area.

On the other hand, extensive sea ice, which increases the distance adult
Emperors have to travel to the edge of the ice pack to feed, leads to reduced
survival of chicks. Nevertheless, in these long-lived birds, the effect of reduced sea ice on adult survival is the most serious influence on their population numbers.

In spite of the decline in extent of sea ice in the East Antarctic region,
the length of the sea ice season there has tended to increase. The reduction
in krill abundance and the later breakup of sea ice has apparently caused
most flying seabirds to arrive at the Dumont d'Urville colony and initiate
egg laying later. For the Southern Fulmar, Antarctic Petrel, Cape Petrel, and
Wilson's Storm-Petrel, these delays ranged from slightly over 5 to more
than 30 days on average over the period 1950–2004. On the other hand,
South Polar Skuas, which prey on the eggs and nestlings of other seabirds,
have tended to initiate nesting a few days earlier.

The breeding populations of Snow Petrels at Terre Adélie are also positively related to the extent of sea ice in their feeding areas. Thus, continued
reduction of sea ice due to climatic warming will almost certainly lead to
continued reduction in their populations.

Also at Terre Adélie, the Southern Fulmar, another Antarctic petrel,
shows highly variable breeding success, correlated with ice conditions. The
Southern Fulmar feeds primarily on krill in areas of pack ice with about 50
percent open water. A 39-year study at Terre Adélie showed that adult survival was low in years with high sea-surface temperatures and pack-ice density lower than 40 percent. Thus, this species is likely to be affected by continued warming of Antarctic waters and recession of sea ice.

In the Ross Sea, where the southernmost colonies of Adélies occur,
populations of Adélie Penguins increased considerably during the 1980s
and have remained at high levels since then. Nonbreeding birds forage in
the 100-kilometer-wide zones bordering the edge of the ice pack. Some
warming is evident in the Ross Sea area, and the extent of sea ice has decreased somewhat. In this area, Adélie populations appear to be favored by

decrease in ice cover, with about a 5-year lag. These changes are still not optimum, and Adélie numbers appear likely to continue to increase with further retreat of the ice cover.

Modeling studies suggest that continued warming of Antarctic ecosystems by 2°C, which will likely occur between AD 2026 and 2060, will severely reduce Adélie and Emperor Penguin populations. Both of these species are limited to areas that have a substantial cover of pack ice for much of the year. Populations of these species north of 70°S, including colonies on the Antarctic Peninsula and much of East Antarctica, will likely decline severely or disappear. On the one hand, Adélie Penguins may be able to colonize new southern areas for breeding where exposed coastline areas appear and pack ice becomes less dense. On the other hand, decreasing areas of ice pack north of the Antarctic Circle, where Adélies now spend the nonbreeding season, may be detrimental to this species.

## Subantarctic Islands

In the South Orkney Islands, three congeneric penguin species occur: the Adélie, Chinstrap, and Gentoo. These three species differ in their preferred foraging areas. The Adélie forages close to the margin of the sea ice, the Chinstrap in open ocean waters, and the Gentoo in inshore waters. Gentoos tend to take more fish, relative to krill, than Adélies and Chinstraps. The South Orkneys, like the Antarctic Peninsula, have experienced gradual warming since the early 1980s and a gradual southward withdrawal of sea ice. The pattern of sea ice around the South Orkneys is also influenced by the Pacific El Niño/La Niña cycle, with a delay of about 15 months relative to the tropical ocean. El Niño conditions lead to warmer sea temperatures, reduced sea ice, and reduced food availability for penguins.

Changing conditions of temperature and sea ice have led to changes in relative numbers of the three penguins in the South Orkneys. Adélies, the most numerous, have experienced a general decline at a rate of about 2.8 percent annually since 1987. Their populations, however, show peaks of 2–3 year duration, corresponding to La Niña periods of more-northward position of the edge of sea ice. Chinstraps have also tended to decline, at an average rate of 1.3 percent annually, and also show increases when sea ice expands northward. Gentoos, in contrast, have increased in numbers steadily, at a rate of about 5.5 percent annually.

On South Georgia, during the 1980s, populations of Gentoo and Macaroni Penguins and Black-browed Albatrosses, all primarily krill feeders, were constant or increasing. Since then, all three species have declined in numbers, as have populations of Antarctic fur seals, which also feed heavily on krill. Although detailed data on krill abundance around South Georgia are lacking, it appears likely that since about 1990, krill abundance has declined. Data on populations, reproductive success, and survival of Black-browed Albatrosses indicate that the population is declining at a rate of about 6.4 percent annually. The decline began about 1976, but its rate has increased since about 1988. At the current rate of decline, modeling indicates that the population could approach extinction in about 78 years.

Several species of penguins breed in the Falkland Islands. The most numerous species are the Gentoo, Rockhopper, and Magellanic. The populations of all three have declined substantially in recent decades. In 2000–2001, Gentoo populations equaled about 113,000 pairs, having recovered from a decline to an estimated 65,000 pairs in 1995–1996. Rockhopper Penguins numbered about 272,000 pairs in 2000–2001, about 11 percent of their numbers in the early 1980s. Populations of Magellanic Penguins, which nest in burrows, are difficult to estimate but are believed to number about 24 percent or less of their original population. Although Gentoo and Rockhopper Penguin populations appear to have stabilized at new, lower levels, populations of Magellanics are still declining.

All three penguin species in the Falklands feed on loligo squid and Blue Whiting, both of which are taken heavily by the local commercial fishing industry. Currently, these species constitute about 5.9 percent of the diet of Gentoos, 10.2 percent of the diet of Rockhoppers, and 26.5 percent of the diet of Magellanics. It is likely that all three penguins relied more heavily on these foods prior to the development of the commercial fishery.

About two-thirds of the world's Black-browed Albatrosses, which feed heavily on krill and krill-eating fish, breed in the Falkland Islands. The population is now just under 400,000 pairs, having declined about 25 percent over the past 20 years. The primary cause of decline here is believed to be mortality of albatrosses associated with commercial longline fishing and trawling. Albatrosses are caught on the longline hooks and drowned when they are dragged underwater. They are also killed when they collide with trawling gear.

In contrast to the situation in the Falklands, populations of Magellanic Penguins in southern Chile, where they are protected from commercial fishing, are stable. Thus, it appears than competition between penguins and commercial fishing underlies most of the population declines in the Falkland Islands.

In the Crozet Archipelago, southeast of South Africa, where about two-thirds of the world's King Penguins breed, warming waters are posing a serious threat to their populations. The rearing cycle of this species lasts 10 months, during which the adults leave the chicks for 5 winter months and return in the spring to feed them until they fledge. The warm phase of the El Niño Southern Oscillation leads to warm waters, which appear to reduce the productivity of phytoplankton and the invertebrates that adult King Penguins feed their young. The distance adults have to swim to reach the food-rich Antarctic Polar Frontal Zone is also increased, which appears to affect the condition of adults and increase their tendency to abandon their young. Survival of adults also appears to be reduced, after a 2-year lag, by high sea-surface temperatures along the northern boundary of the pack ice (Figure 16.1). This appears to be a result of the effects of warm waters on populations of prey on which adults depend.

Populations of five species of albatrosses and giant petrels nest in the Crozet Archipelago, too. The albatrosses are subject to mortality due to the longline fishery, but full information of these losses is difficult to obtain, and the effect on their populations is uncertain. The Sooty Albatross has shown a continuous decline since the 1980s, correlated with warming of the Southern Ocean, and exhibits a decrease in breeding pairs during El Niño years.

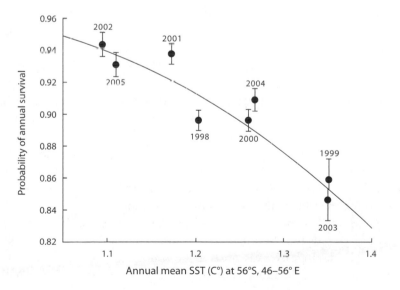

FIGURE 16.1. Probability of survival of adult King Penguins in the Crozet Islands in relation to mean annual sea surface temperature (SST) 2 years earlier. (From Le Bohec et al. 2008.)

Populations of several species of penguins on Marion Island and the Prince Edward Islands, in subantarctic waters southeast of the southern tip of Africa, are declining. King, Gentoo, Rockhopper, and Macaroni Penguins all breed on the islands. It is possible that a southward shift in the Antarctic Circumpolar Current has displaced food-rich waters southward beyond the foraging range of these penguins. On Marion Island, populations of several albatrosses, Southern and Northern Giant-Petrels, and the White-chinned Petrel have also declined since the late 1990s. These declines seem to be largely the effect of bycatch by the longline fishery.

## Temperate and Tropical Regions

In the Galápagos Islands, nine El Niños between 1965 and 2004 have affected the Galápagos Penguin population. Two of these events were unusually severe, compared with previous events. The population declined by about 77 percent following the strong El Niño in 1982–1983 and by about 65 percent following the strong El Niño in 1997–1998. During less severe El Niños, adult penguins appear able to survive, but little or no recruitment of young birds occurs. The overall frequency of El Niños in recent years is several times greater than in the past, and that has impeded population recovery from severe El Niños. In 2004, the estimated penguin population was about 1500 birds, a little over a third of the population estimated to exist in 1971.

The long-term prospects for survival of the Galápagos Penguin are uncertain. A population viability analysis carried out by the Charles Darwin Foundation and the Galápagos National Park Service in 2005 estimated that a 30 percent chance exists that the species will become extinct in the next century because of the effects of intensified El Niños. Predation by feral cats has also become a problem on Isabela Island. The Humboldt Penguins of coastal Peru and Chile are also affected by El Niños, as well as fishing activities. Declines in Magellanic Penguins in southern Chile and Argentina may also be partly due to climatic warming.

The Sooty Shearwater, one of the most numerous seabirds in the world, breeds on islands near South America, Australia, and New Zealand. Nonbreeding birds range over vast areas of the South and North Pacific, some of them flying about 72,420 kilometers in the course of a year. Their movements appear to be timed to enable them to exploit ocean areas at times of their highest productivity in shearwater foods. Their populations are declining both at breeding colonies near New Zealand and in nonbreeding areas of the North Pacific. The reasons for these declines are uncertain at present but may be related to changing climate and prey availability.

In the western tropical Pacific, abnormally high sea-surface temperatures in 2002 at Heron Island in the southern Great Barrier Reef area led to almost complete failure of reproduction by Wedge-tailed Shearwaters. Reduced rates of feeding by adults led to heavy mortality of the young. This, in turn, was due to the fact that foraging success of shearwaters was adversely affected by high sea temperatures. Sea temperatures were high enough to cause severe coral bleaching, as well. At Michaelmas Cay, over the period 1984 to 2001in the northern Great Barrier Reef, Sooty Terns and Brown Noddy terns showed strong reductions in breeding populations correlated with El Niños. In this case, the birds responded to precursor indicators of El Niño conditions that did not develop fully until the following year. These observations suggest that additional responses of tropical seabird populations due to climatic change are very likely.

# Summary

Warming of the southern oceans and reduction in area of pack ice have led to a major reduction in productivity of krill, the mainstay of food chains leading to vertebrates. The greatest population changes have occurred along the Antarctic Peninsula, where warming and change in the pack ice regime are greatest. Some penguin species have benefitted and others have suffered. In the Ross Sea and elsewhere, Emperor Penguins have declined to about half their original numbers, but Adélie Penguin populations have increased. Southern Fulmars and several petrel populations have also declined. On various subantarctic islands, penguins of several species have generally tended to decline in numbers, as have other seabirds, such as the Black-browed Albatross. In tropical waters, the Galápagos Penguin has declined with the increasing frequency of El Niños, and some shearwaters appear to be suffering population declines with ocean warming.

### KEY REFERENCES

Ainley, D., J. Russell, and S. Jenouvrier. 2008. The fate of Antarctic penguins when Earth's tropospheric temperature reaches 2°C above pre-industrial levels. World Wildlife Fund. http://assets.panda.org/downloads/wwf_climate_penguins_final.pdf.

Arnold, J. M., S. Brault, and J. P. Croxall. 2006. "Albatross populations in peril: A population trajectory for Black-browed Albatrosses at South Georgia." *Ecological Applications* 16:419–432.

Atkinson, A., V. Siegel, E. Pakhomov, and P. Rothery. 2004. "Long-term decline in krill stock and increase in salps within the southern ocean." *Nature* 432:100–103.

Barbraud, C. and H. Weimerskirch. 2006. "Antarctic birds breed later in response to climate change." *Proceedings of the National Academy of Sciences USA* 103:6248–6251.

Boersma, P. D. 2008. "Penguins as marine sentinels." *BioScience* 58:597–607.

Croxall, J. P., P. N. Trathan, and E. J. Murphy. 2002. "Environmental change and Antarctic seabird populations." *Science* 297:1510–1514.

Devney, C. A., M. Short, and B. C. Congdon. 2009. "Sensitivity of tropical seabirds to El Niño precursors." *Ecology* 90:1175–1183.

Forcada, J., P. N. Trathan, K. Reid, E. J. Murphy, and J. P. Croxall. 2006. "Contrasting population changes in sympatric penguin species in association with climate warming." *Global Change Biology* 12:411–423.

Le Bohec, C., J. M. Durant, M. Gauthier-Clerc, N. C. Stenseth, Y. H. Park, R. Pradel, D. Grémillet, J.-P. Gendner, and Y. Le Maho. 2008. "King penguin population threatened by Southern Ocean warming." *Proceedings of the National Academy of Sciences USA* 105:2493–2497.

Smith, R. C., D. Ainley, K. Baker, E. Domack, S. Emslie, B. Fraser, J. Kennett, et al. 1999. "Marine ecosystem sensitivity to climate change." *BioScience* 49:393–404.

Vargas, H., C. Lougheed, and H. Snell. 2005. "Population size and trends of the Galápagos Penguin *Spheniscus mendiculus.*" *Ibis* 147:367–374.

# PART IV

## Evolutionary Responses of Migratory Birds

We now know that evolutionary change underlies many of the adaptive responses of migratory birds but also that we are probably just seeing the tip of the evolutionary iceberg. Evolutionary changes in morphology, physiology, and behavior related to migration are best documented for small land birds that can be studied experimentally, but some larger land birds and waterbirds show changes in migration behavior that likely have an evolutionary basis.

# Chapter 17

# *Land Birds: Evolutionary Adaptability*

The behavior and physiology of migratory birds help them respond to changes in weather from year to year, but these capabilities alone may not enable the species to adapt to rapid and progressive climatic change. Many responses of migratory land birds to changing climate have an evolutionary basis. These include changes in breeding and nonbreeding ranges, directions and distances of migration flights, timing of spring and fall migration, proportions of migrating and resident individuals, and other hereditary characteristics related to migration.

Warming climates at high latitudes now favor earlier breeding by many species. Thus, genetic variability in the timing of migration between nonbreeding and breeding areas, and in the conditions triggering departure from nonbreeding areas, become critical. This variability is especially important for long-distance migrants. Whether or not these species can evolve fast enough to keep up with the rapid changes in their environment is uncertain. Genetic linkages between life cycle processes such as breeding, molt, and migration can constrain the ability of species to alter the timing of migration through evolution.

## Range Changes

Breeding and nonbreeding ranges of land birds have changed in the past, in most cases due to modification of habitats by human activities. As we have

seen in earlier chapters, however, climatic change is also responsible for altering ranges. Some range shifts may simply reflect the intrinsic adaptability
of individuals in their migratory movements and in their selection of breeding or nonbreeding sites. In other cases, the changes of migration direction
or distance that lead to occupation of new ranges apparently have an evolutionary basis.

## Breeding Range

An interesting example of a recent range shift is provided by the Dark-eyed
Junco in southern California. When I was living in San Diego in 1983, this
species established a breeding population on the La Jolla campus of the
University of California, San Diego (UCSD). The campus is a landscaped
woodland with pines, eucalyptus, and other evergreen trees and enjoys a
mild Mediterranean climate. The environment is quite different from the
mixed conifer and pine–oak forests of the Palomar and Cuyamaca mountains in eastern and northern San Diego County, where the Dark-eyed
Junco is a common breeding bird, and from where the campus birds probably came. These mountain birds are common winter visitors to the coastal
areas of the county, including the UCSD campus. The birds nesting on the
campus, however, differ from those in the mountains in being permanent
residents, instead of altitudinal migrants, and in having a much longer
breeding season. The campus birds produce as many as four broods each
year. They also show a weaker tendency to flock. Although a genetic basis
for all of these changes is not certain, the campus population shows genetic
differences from mountain populations in plumage, as we shall see below.
This suggests that adaptation to the coastal environment is at least partly
through evolutionary change.

## Nonbreeding Range

Perhaps the best-documented example of an evolutionary shift in nonbreeding range is that of the Blackcap population wintering in the British
Isles. One of the most familiar and widespread Old World warblers, the
Blackcap breeds from the British Isles east to Siberia and south as far as the
mountains of North Africa. Since the early 1980s, Peter Berthold and other
scientists at the Max Planck Institute for Ornithology in Germany have
used this species to investigate the genetics and evolutionary biology of

bird migration through laboratory and field studies. The enormous success of their work has stimulated similar investigations in many other laboratories in Europe and North America.

Prior to the 1960s, western European populations of the Blackcap wintered in the Iberian Peninsula, North Africa, and sub-Saharan western Africa. Eastern European warblers flew south to winter in East Africa. Beginning in the 1960s, however, Blackcaps began wintering in large numbers in England and Ireland. Surprisingly, rather than being British breeding birds that did not migrate south during the winter, these birds came from central Europe, having migrated west-northwest rather than south in autumn. The tendency for central European birds to use this new wintering area increased rapidly over three decades, and by about 1990, from 7 to 11 percent of the breeding birds in Germany and Austria, and about 35 percent of those in Belgium, wintered in the British Isles.

The compass direction of migration by Blackcaps wintering in the British Isles is 45 to 90° different from the direction to their former wintering area. Captive breeding of Blackcaps eventually showed that the change in migratory orientation had a genetic basis. Blackcaps trapped in the British Isles in winter, and later tested in orientation chambers during the time of fall migration, oriented appropriately for migration to the British Isles. Offspring of these birds also showed the same orientation pattern. Also as expected of a genetic adaptation, hybrid offspring of these birds and mates obtained in central Europe showed an orientation pattern intermediate between that of birds wintering in the British Isles and those wintering in southwestern Europe.

What has favored this evolutionary change? The present suitability of the British Isles as a wintering area may, in part, be due to the milder winter climate that now characterizes the region. It is also possible that human activity or climatic change has degraded parts of the traditional winter range in Africa.

Recently, too, Blackcaps wintering in Britain and Ireland were discovered to migrate to their breeding range in Germany and Austria about 2 weeks earlier than those wintering in Spain and Portugal. This earlier migration may reflect the influence of increasing day length in spring on earlier gonadal development and migratory readiness of birds wintering in more-northern areas. Increasing day length is known to be a strong stimulus to these changes in birds of the North Temperate Zone. For birds spending the winter farther north, the increase in day length after the spring equinox is faster that at locations farther south. The result of earlier migration to breeding areas is that Blackcaps wintering in the British Isles are about two and a half times more likely to mate with each other than with Blackcaps from other wintering locations. This pattern of selective mating is probably reinforcing the evolutionary divergence of Blackcaps wintering in the British Isles.

As suggested by the Blackcap example, shifts of wintering areas may help adapt long-distance migrants to changing conditions on their breeding ranges. Experimental studies with trans-Saharan migrants such as Garden Warblers, Common Redstarts, and European Pied Flycatchers support this idea. Birds of these species were held during the winter under day lengths simulating a wintering area north of the Sahara. In this case, day lengths shorter than those of their normal sub-Saharan winter range appear to bring the birds into migratory readiness earlier in the season. Thus, a genetic tendency for individuals for these species to winter north of the Sahara might be favored strongly by an accompanying tendency to migrate earlier and arrive earlier on the breeding range. Other studies suggest that the Great Reed-Warbler may have begun to exhibit such a range change. Some individuals of this species, which normally winters in sub-Saharan Africa, are now wintering in Spain.

In North America, a similar case may be presented by the Rufous Hummingbird. This hummingbird breeds in the Pacific Northwest, from Oregon and Idaho north to southern Alaska. Where I live in northern New Mexico, we know it as the "Mean Old Rufous" that dominates all other hummingbirds at our feeders when the males arrive in midsummer on their way south. Its traditional wintering range extends from northwestern Mexico southward. Although it was noted sporadically in the southeastern United States in the early 1900s, wintering birds began to become common along the Gulf of Mexico coast from Louisiana eastward in the 1970s. In the 1980s and 1990s, the number of wintering Rufous Hummingbirds has steadily grown, and they now winter eastward through Mississippi, Alabama, and Georgia to Florida and South Carolina. Wintering birds now occur up to 400 miles north of the Gulf coast in places such as Atlanta, Georgia.

The first Rufous Hummingbirds to appear in the Southeast are usually seen close to the Gulf coast in July and August. Later, in November and December, wintering birds appear at localities farther east and inland. Although experimental studies have not yet demonstrated that this new wintering area and migration pattern have a genetic basis, the situation is remarkably similar to that of the European Blackcap. The selective basis for such an evolutionary change would seem to be the milder winter climate that has developed along the Gulf coast, together with the increase in nectar-producing ornamental plants and in hummingbird feeders due to human activity. It is also possible that the suitability of traditional wintering areas in Mexico has declined.

## Timing of Migration

The timing of spring migration by birds breeding in the North Temperate Zone appears to be under strong genetic control for most small land birds. In North America, a clear example of natural selection for a change in timing of spring migration is provided by the Cliff Swallow. This species is colonial, building tubular mud nests on natural cliffs, under bridges, and on buildings. In spring the swallows often return to colony sites in large numbers over a few days. When a severe cold snap followed the arrival of the first swallows at a colony in Nebraska in 1996, a large fraction of the birds that had arrived early were killed. In subsequent years, the arrival of birds at the colony in question was delayed significantly, indicating that selection had favored birds with a genetic tendency for later migration.

In the Old World, the earlier arrival by trans-Saharan migrant birds in southern Europe raises the question of whether earlier arrival reflects a change of genetic timing of northward migration or just the response of birds to changes in sub-Saharan weather. Niclas Jonzén, a Swedish investigator, and his colleagues in several other European laboratories have documented earlier arrival of nine species of trans-Saharan migrants in southern Italy and in four locations in Scandinavia. The wintering area of these species has warmed, which could be a cue triggering earlier departure on northward migration. But, if anything, the warmer and drier weather is also likely to reduce the food available to these birds, which should delay the start of their migration. The latter possibility led these authors to suggest that the earlier migration schedule of these species is probably a climate-driven evolutionary change based on the advantage of earlier arrival on the breeding range.

Other workers have examined this issue by analyzing the spring arrival of the Bank Swallow in the United Kingdom. Between 1950 and 2005, both the Bank Swallow and the Barn Swallow have tended to arrive earlier, the Bank Swallow by roughly 20 days. In addition, the correlation between Bank Swallow arrival and temperature has changed. Throughout the years, warmer springs have led to earlier arrivals. Between 1978 and 2005, however, the arrival date changed much faster with a given increase in temperature than it did between 1950 and 1977. For the Barn Swallow, on the other hand, the relationship of arrival date to temperature has remained constant. Both swallows winter in sub-Saharan Africa, and their migration routes overlap north of the equator. The pattern for the Bank Swallow suggests that an evolutionary change has occurred, allowing it to advance its arrival on the breeding grounds faster than in the period from 1950 to 1977.

Although the Barn Swallow did not show the same pattern as the Bank Swallow in the United Kingdom, studies in mainland Europe have shown that its arrival date also has a genetic basis. In addition, Barn Swallows that arrive early in spring also tend to have longer tails, a characteristic that seems to aid in their ability to maneuver and capture insects in flight. This ability seems to be at a premium in early spring, when flying insects are likely to be scarce.

Factors controlling the fall migrations of land birds have received less attention. Studies have revealed, however, that the timing of fall migration in Blackcaps is also under genetic control. In laboratory experiments, for example, the onset of fall migratory activity could be delayed by 2 weeks through selection over just two generations. The timing of fall migration in the Garden Warbler, another European long-distance migrant, is also known to be under strong genetic control.

A few studies have attempted to determine whether these patterns of evolutionary change in timing of migration are widespread among land birds. One such effort examined patterns of advance in migration timing of more than twenty European land birds in a meta-analysis designed to detect patterns attributable to evolutionary change. This study looked for evidence of more-rapid change in timing of migration in birds with earlier age at first reproduction, and thus shorter generation times. The assumption of this test is that species with shorter generation times can show faster evolutionary change. The analysis confirmed that earlier migrations by all species were correlated with climatic warming. Comparison of rates of change in spring arrival times of birds with different generation lengths, although consistent with a certain rate of evolutionary change, were also consistent with strictly phenotypic responses. Thus, the analysis provided no firm evidence for or against evolutionary change.

## Migratory versus Resident Populations

Changing climate is leading to changes in the proportion of migrants and residents, both in bird communities at large and within individual species. In the Lake Constance area of central Europe, for example, the relative abundance of long-distance migrant species declined and that of short-distance migrants and residents increased between 1980–1981 and 1990–1992. Over this period the temperature of the coldest month increased by 2.0°C.

In addition, populations of partial migrants are changing to residents, and short-distance migrants to partial migrants, in some areas. In central Europe, several species that were formerly short-distance migrants have now begun to remain throughout the winter. In Germany, many species of short-distance and partial migrants are showing an increase in the fraction of birds migrating less than 100 kilometers to wintering sites.

Banding recoveries also show changes in the migration strategies of two European thrushes, the Redwing and the Eurasian Blackbird, between 1970 and 1999. For the Eurasian Blackbird, a partial migrant in France, Switzerland, and southern Germany, the fraction of the population that was migratory declined continuously during the 30-year period. This pattern agrees with the prediction based on genetic studies with other partial migrants, such as the Blackcap. For the Redwing, a complete migrant, recoveries indicated a shortening of migration distance. Migration distance in a given year was strongly related to the severity of the winter before, suggesting that the response of this species was simply a behavioral response, rather than a genetic change.

## Morphology

Several aspects of body structure seem to be influenced by changing climate, with evolutionary implications. In Israel, where minimum summer temperatures have warmed 0.26°C per decade since about 1950, the body mass of the Sardinian Warbler, a partial migrant, has declined about 27 percent. The body mass of several other resident passerines has also declined. Smaller mass and a greater surface-to-mass ratio may facilitate heat loss by these species under a warmer climate. In England, where the climatic has warmed appreciably, the body mass of Eurasian Reed-Warblers and Black-caps declined over the period between 1968 and 2003. These same species also showed increases in wing length. In Germany, a decline in body mass of the Common Chiffchaff, another Old World warbler, was also noted over the period 1981–2003. These changes may also reflect, at least partially, evolutionary responses to climatic warming.

The Barn Swallow has proven to be an interesting subject for evolutionary studies of change in body structure. One of its secondary sexual characteristics is the differentially longer outer tail feathers in the male, compared with those of the female. Males with well-developed outer tail feathers achieve greater reproductive success, and other evidence suggests that birds with longer outer tail feathers are healthier and perhaps more efficient at feeding on flying insects. Barn Swallows breeding in Denmark migrate from their winter home in southern Africa through the Iberian Peninsula. Over the period between 1984 and 2003, males in the Danish breeding population showed an increase in the average length of the outer tail feathers. Analysis of annual survival rates of males in relation to geographical and weather indices correlated with the North Atlantic Oscillation revealed that the northern African area centered on Algeria was critical to the survival of nonbreeding birds. Over the period of study, a satellite-based index of vegetation "greenness," termed the Normalized Difference Vegetation Index, has declined progressively in Algeria, reflecting a trend toward drier conditions and reduced abundance of flying insects. These drier conditions at the jumping-off point for northward flights into Europe may be a strong selective agent against males in poor physical condition, reflected in their weaker development of elongated outer tail feathers. If these birds are less efficient feeders, their fat stores may be inadequate for flights across the Mediterranean to Europe.

Severe weather events that inhibit feeding can cause severe mortality for species such as Barn Swallows. In North America, a 6-day period of cold, rainy May weather caused heavy mortality of Barn Swallows in Nebraska. Comparison of birds that died with those that survived showed that survivors had larger bodies, longer bills, and more symmetric wings and tails. These wing and tail features are thought to enable the birds to forage more efficiently during bad weather. Larger size is probably also an indication of greater fat reserves. Somewhat similar results were found for Cliff Swallows after the same weather event. These episodes simply demonstrate that natural selection by climate-related events is ongoing.

Another quite different morphological feature, egg size, may also be related to climate change. The size of eggs of the European Pied Flycatchers nesting in northern Finland increased over the period from 1975 to 1993. This increase was positively correlated with increasing temperatures during the egg-laying period and also with increasing hatching success. Whether this represents a genetic change or a phenotypic response somehow related to warmer weather is uncertain.

In the Dark-eyed Junco population that has become established in coastal San Diego County in southern California, plumage differences have also appeared. The amount of white on the sides of the tail is about 17 to 20

TABLE 17.1. Percentage of white feathers in the tails of Dark-eyed Juncos in San Diego County populations.

|                          | Males | (Number) | Females | (Number) |
|--------------------------|-------|----------|---------|----------|
| Mountain breeding birds  | 45.1  | 14       | —       | —        |
| Coastal wintering birds  | 43.9  | 18       | 38.1    | 16       |
| UCSD resident birds      | 36.3  | 104      | 31.8    | 112      |

*Source:* Data from Yeh 2004.

percent less than in birds in the mountain populations from which they are apparently derived (Table 17.1). Hand-raised birds from both coastal and mountain populations retain the difference in tail pattern when raised under the same conditions, indicating that the difference is hereditary. The rate of evolutionary change in this characteristic was estimated to be 0.19 haldanes, or about 0.4 standard deviations, per generation. Although many environmental differences are associated with the change in habitat and loss of migratory behavior by these juncos, little doubt exists that these changes reflect rapid evolution.

## Evolutionary Adaptability

Several lines of evidence suggest that many birds possess enough genetic variability to enable them to adapt to changing climatic conditions through natural selection. In North America, the House Finch is providing a remarkable example of this adaptability. House Finches are native to western North America, where they are partial migrants. In California, for example, only about 2 to 3 percent of individuals are migratory. House Finches throughout eastern North America are derived from cage birds that were released on Long Island, New York, in about 1940. This release reputedly followed passage of a federal law prohibiting possession and sale of native migratory birds. From this introduction, in any case, the birds have spread throughout the East from southern Canada to Florida and westward until they have met their western relatives. In the more seasonal eastern environment, migratory behavior has become more pronounced, with 28 to 54 percent of individuals now migrating (Table 17.2). In addition, the average distance of their migratory movements has increased over the years.

Furthermore, the wing shape of eastern North American House Finches now differs from that of western birds. In a comparison of birds

TABLE 17.2. Percentage of migratory House Finches in long-established populations in California and recently established populations in the eastern United States.

| Geographical Area | Years | Number of Birds | Percent Migratory[1] |
|---|---|---|---|
| Source Population | | | |
| Southern California | 1992–1997 | 460 | 2.1 |
| Northeastern Core Population[2] | | | |
| | 1978–1984 | 52 | 31.5 |
| | 1988–1992 | 26 | 43.0 |
| Eastern Expansion Populations | | | |
| Western Pennsylvania | 1978–1984 | 52 | 46.1 |
| Ohio | 1981–1985 | 20 | 50.0 |
| Indiana | 1988–1992 | 26 | 76.9 |
| Southeastern United States | 1988–1994 | 19 | 42.1 |

Source: Data from Able, K. P. and J. R. Belthoff. 1998. "Rapid 'evolution' of migratory behavior in the introduced House Finch of Eastern North America." Proceedings of the Royal Society of London B 265:2063–2071.
[1] Banded birds recovered > 80 km from banding locality
[2] New York and Pennsylvania east of 76° longitude, New England and mid-Atlantic states

from a population near Boise, Idaho, with those from Ithaca, New York, the eastern birds had more-pointed wings, due primarily to the fact that their proximal primaries were significantly shorter than those of western birds. In addition, the most distal primary wing feather of eastern birds was significantly longer than that of western birds. The more pointed shape of the wings of eastern birds is very similar to that of other migratory birds, in contrast to close relatives that are permanent residents.

The Great Tit presents an interesting case study of how adaptation to climate change can vary within a species. This Old World species is a permanent resident in western and southern Europe but a partial or altitudinal migrant in northern Europe and eastward through eastern Europe and Asia. With climatic warming in western Europe, caterpillars on which these birds specialize during nesting are reaching their peak abundance earlier in the season. In the Netherlands, overall nesting by Great Tits has not advanced to keep pace with caterpillar availability. Females did nest earlier in warm springs, but individual birds varied substantially in their response. This variation has a genetic component, and selection appears to favor females that advance their laying dates in response to warm temperatures.

The synchrony of laying time with food peaks over the lifetime of females is greater in females that show a capacity to advance laying dates in warm springs. So, it appears that a degree of evolutionary adjustment is occurring. A number of other studies in the Netherlands have found a high degree of heritability in foraging behavior of Great Tits.

In England, in contrast, the laying dates of Great Tits have advanced about 14 days over a 47-year period, keeping close synchrony with the spring peak of biomass of moth caterpillars. Data on the laying behavior of individual females shows that they are very responsive to spring weather conditions, advancing or delaying nesting depending on weather, but that females do not vary significantly in this response. Thus, individual adaptability seems adequate to account for the population response. Lack of genetic variability in response to spring weather in these English birds seems somewhat surprising in view of observations in mainland Europe.

## Genetic Analysis

Genetic studies are now beginning to look for the exact genetic basis for some of the evolutionary differences that are being recognized in migratory birds. One study, for example, has investigated enzyme patterns in Blackcap populations that differ in the proportion of migratory individuals. In this study, the genetic loci for fourteen enzymes were examined, one of which proved to be strongly associated with migratory behavior. A specific form of one of the enzymes was found only in populations with individuals that did not migrate. In addition, this form of the enzyme increased in frequency in experimental populations that were subjected to selection to become sedentary. This is obviously just the first step in examining the molecular basis for evolutionary differences in migration patterns.

## Summary

Many examples of evolutionary adjustments of morphology, physiology, and behavior are now documented for land birds. Shifts of breeding and nonbreeding ranges have occurred. Shifts of nonbreeding areas, coupled with responses to new photoperiod regimes, may be adaptive in triggering spring migration earlier in the season. Changes have also been observed in the proportions of resident and migrant individuals in populations, as well as in several morphological features, including body and egg mass, characteristics of flight feathers, and secondary sexual features. Genetic variability in response to spring weather may also help birds adjust breeding schedules to food availability. The timing of both spring and fall migration has been shown to be under strong genetic control. Genetic analyses are now beginning to identify the precise loci involved in adaptation by migratory species.

## KEY REFERENCES

Bearhop, S., W. Fiedler, R. W. Furness, S. C. Votier, S. Waldron, J. Newton, G. J. Bowden, P. Berthold, and K. Farnsworth. 2005. "Assortative mating as a mechanism for rapid evolution of a migratory divide." *Science* 310:502–504.

Coppack, T., F. Pulido, M. Czisch, D. P. Auer, and P. Berthold. 2003. "Photoperiodic response may facilitate adaptation to climatic change in long-distance migratory birds." *Proceedings of the Royal Society of London B* (Supplement) 270:S43–S46.

Egbert, J. R. and J. R. Berthoff. 2003. "Wing shape in house finches differs relative to migratory habit in eastern and western North America." *Condor* 105:825–829.

Gienapp, P., R. Leimu, and J. Merilä. 2007. "Responses to climate change in avian migration time: Microevolution versus phenotypic plasticity." *Climate Research* 35:25–35.

Helbig, A. J. 1996. "Genetic basis, mode of inheritance and evolutionary changes of migratory directions in Palearctic warblers (Aves: Sylviidae)." *Journal of Experimental Biology* 199:49–55.

Jonzén, N., A. Lindén, T. Ergon, E. Knudsen, J. O. Vik, D. Rubolini, D. Piacentini, et al. 2006. "Rapid advance of spring arrival dates in long-distance migratory birds." *Science* 312:1959–1961.

Lemoine, N. and K. Böhning-Gaese. 2003. "Potential impact of global change on species richness of long-distance migrants." *Conservation Biology* 17:577–586.

Møller, A. P. and T. Szép. 2005. "Rapid evolutionary change in a secondary sexual character linked to climatic change." *Journal of Evolutionary Biology* 18:481–495.

Sparks, T. and P. Tryjanowski. 2007. "Patterns of spring arrival dates differ in two hirundines." *Climate Research* 35:159–164.

Yeh, P. J. 2004. "Rapid evolution of a sexually selected trait following population establishment in a novel habitat." *Evolution* 58:166–174.

# Chapter 18

# *Waterbirds: Evolutionary Adaptability*

Birds associated with freshwater and coastal marine aquatic habitats are likely to be particularly at risk from climate change as temperature and precipitation patterns are modified over large areas. Rising sea levels will also reduce the area of habitat of these birds in some coastal marine areas. Influences on waterfowl are especially difficult to predict because of the complicating influences of hunting harvests and change in agricultural cropland distributions.

Detecting evolutionary responses in these birds, many of which are large and long-lived, is much more difficult than for small land birds. The rate of evolutionary adaptation is inversely related to generation time, which is longer for most large waterbirds than for small land birds. However, a number of studies demonstrate or strongly suggest that such responses do exist and are likely to contribute to the overall response of waterbirds to changing climate. In this chapter we shall review the few available studies that demonstrate or suggest evolutionary responses by freshwater and marine birds.

## Grebes

From the 1950s through the early 1990s, the population of Great Crested Grebes, a partially migrant species, increased rapidly throughout much of

Europe. The proportion of grebes wintering in the breeding area in the Netherlands also increased from 22 percent prior to 1970 to over 80 percent after 1980. This was correlated with a decrease in the number of grebes wintering farther south in Switzerland. Evidence relating to the timing of changes in wintering patterns and of the pattern of migration of grebes hatched at different times during the breeding season argued that the increase in residency was a consequence of evolutionary change. This is interpreted as natural selection favoring birds that overwinter closer to breeding sites and also favoring nonmigratory individuals. The selection response in both traits holds the potential for creating rapid evolutionary change from partial migration to complete residency. In this case, it appears likely that increasing availability of artificial lakes, rather than climatic change alone, may have contributed to the evolutionary change.

## Shorebirds

Sandpipers of the subgenus Calidrinae present an interesting pattern of migration in relation to habitat use and possibly to genetic variability. Arctic-breeding species tend to be long-distance migrants and tend to winter in coastal marine environments. The Red Knot, breeding in the high Arctic, has low genetic variability as a consequence of a severe late Pleistocene genetic bottleneck. Other high-Arctic breeders, including the Curlew Sandpiper, also exhibit low genetic variability, likely also the result of genetic bottlenecking when their ranges were greatly restricted during Pleistocene glaciations. The Dunlin, on the other hand, breeds in the low Arctic and North Temperate Zone and shows high genetic variability.

Other shorebirds that exhibit low genetic variability include two long-distance migrants, Black-bellied Plover and Willet, and a short-distance migrant, the Mountain Plover. On the other hand, the Ruff, a long-distance migrant that breeds in southern Europe in freshwater habitats, exhibits high genetic variability. Thus, for some of the longest-distance migrant shorebirds breeding in the Arctic, genetic variability that may be important in adaptation to the rapidly changing modern climate may be deficient.

On the other hand, the Red Knot consists of migratory subpopulations that winter in the southeastern United States, in Brazil and the Caribbean, and in southern South America. Data from 2004 and 2005 also suggest some segregation in these groups at stopover sites within Delaware Bay versus on the Atlantic coast. These striking differences in migratory pattern might well have a genetic basis. Whether these migratory groups remain segregated or mix on the breeding range is unknown.

Many of the shorebirds that nest in the Arctic exhibit very long migrations to wintering areas in the Southern Hemisphere. Bar-tailed Godwits migrate south from eastern Asia and Alaska, stopping to refuel in the Yellow Sea region of China and Korea and continuing south to New Zealand and eastern Australia. Observation of departure dates of banded birds from New Zealand show that although departure of the entire wintering population spans about 25 days, departures of individuals are much more consistent (Figure 18.1). Of fifty-seven birds, thirty-two departed within the same 7-day period in different years. Among adult birds, twenty-nine of forty-six showed consistent departures within the same week. In twenty-three cases, departures were within a 5-day period. High synchrony of arrival of mated pairs also characterizes Bar-tailed Godwits at breeding areas in Iceland in spite of their having wintered in very different areas. These regularities suggest strong genetic control of the timing of migration, although with substantial variability among individuals.

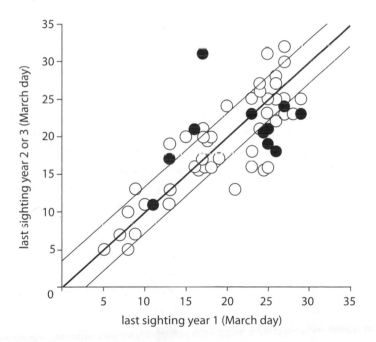

FIGURE 18.1. Departure dates of individual Bar-tailed Godwits from New Zealand in successive years between 2004 and 2006. The center line represents an exact correspondence of departure dates in different years, and the bordering lines represent a difference of 3 days in departure in different years. (From Battley, P. F. 2006. "Consistent annual schedules in a migratory shorebird." *Biology Letters* 2:518. Permission of the Royal Society.)

In the Netherlands, the Northern Lapwing advanced its egg-laying date about 10 days over the period 1897–2003. Most of the advance has occurred since the early 1950s, and appears to be correlated with warming of spring temperatures and increase in winter rainfall. About 30 percent of this advance, however, may be related to changes in farming practices, particularly intensified meadow mowing activities. Whether the advance correlated with temperature and rainfall is entirely a phenotypic response to weather and food availability or whether it is the result of selection related to reproductive success is uncertain. The advance in egg-laying date additional to that related to weather, however, seems likely to be related to selection for earlier nesting. Black-tailed Godwits also advanced their nesting dates in the Netherlands by about 20 days between 1910 and 1975, perhaps for similar reasons. Interestingly, Northern Lapwings in the United Kingdom do not seem to have advanced their nesting dates over the latter part of the past century.

## Waterfowl

A number of studies of waterfowl suggest that significant genetic variability in migratory tendencies exists and that changing habitat conditions may lead to evolutionary adaptation. One of the studied species is the North American Snow Goose. Midcontinent populations of the Snow Goose have grown rapidly since the 1960s. The large populations of these birds, which breed primarily in the James Bay, Hudson Bay, and Foxe Basin regions of Canada, have caused major damage to salt marsh habitats in which they nest. A major contributor to their population growth was the increase in grain production in wintering areas throughout the midcontinent region of North America.

Originally, however, midcontinent Snow Geese wintered largely in coastal salt marshes of Texas and Louisiana, where their feeding activity was similar to that in Canada, involving the excavation of roots and rhizomes of marsh plants. With the development of grain farming, especially rice, in the Gulf States, Snow Geese expanded their winter feeding activities into grain fields. In the 1970s, populations began to winter much farther north, in the central Great Plains where cornfields with unharvested waste grain provided them with abundant food. Wintering birds tend to be faithful to specific winter habitats: salt marsh, rice fields, and cornfields. Furthermore, the birds that favor these specific habitats tend to show characteristic morphologies. Geese wintering in salt marshes tend to be largest in body size and possess the thickest bills and longest culmens, features that can be interpreted as adaptive in excavation of marsh foods. Birds feeding in rice fields are of intermediate size, and those wintering in cornfields are smallest. Bill thickness is least in birds wintering in rice fields, and culmen length shortest in birds wintering in cornfields. The degree to which these differences among wintering populations of Snow Geese are genetic is as yet uncertain. Snow Geese show high genetic similarity across North America. The population on Wrangel Island, north of the Siberian mainland, however, shows genetically distinct subpopulations that winter in separate areas in western Canada and the United States.

The consistency of speed of migration and arrival timing of Snow Geese suggests that these characteristics have a strong genetic basis. Radio tracking of Snow Geese breeding on Bylot Island in the high Canadian Arctic and wintering along the mid-Atlantic coast revealed that migratory schedules have a significant degree of genetic control. Tracking of 75 individuals from wintering to breeding areas for up to 3 years revealed that some individuals consistently showed fast migration and early arrival at the nesting area. This suggests that evolutionary change in migration and arrival could be important in maintaining synchrony between the reproductive schedules of Snow Geese and changing climatic conditions.

The Canada Goose and Cackling Goose complex shows high geographic variability, with populations differing in body size, plumage characteristics, migratory patterns, and genetic structure. Most of these differences are not of recent origin. Two recent colonization events have occurred, however. Middleton Island, in the Gulf of Alaska, was colonized in the early 1980s. Genetic analyses show that these birds share a distinctive mitochondrial DNA haplotype with those on Green Island in Prince William Sound, Alaska. Canada Geese also colonized western Greenland at latitudes 66–71 N sometime in the mid-1900s but bred there in only small numbers until the 1980s. Since then, the population has increased to several thousand birds. Birds captured and measured in 1992 showed similarities to the subspecies that breeds in northern Quebec, Canada, although perhaps showing somewhat longer tarsi. Detailed genetic analyses of birds obtained in 1999 show that the Greenland birds are most similar to geese from the eastern Ungava Peninsula, Quebec. Genetic diversity of the Greenland birds is lower than for birds from the Ungava Peninsula, and Greenland birds show small but significant differences in microsatellite allele frequencies and mitochondrial DNA haplotype frequencies. These differences probably represent only a founder effect. Nevertheless, these newly established populations, especially that in Greenland, possess some degree of geographic isolation from source populations and might be at an early stage of evolutionary divergence.

In a somewhat artificial situation, selection seems to have influenced the reproductive pattern of the Mute Swan, a partial migrant in northern Europe. In a colony in Dorset, southern England, which has existed since the mid-fourteenth century, reproduction has been monitored for 25 years. This population, although free ranging, receives supplementary food and a degree of special protection of cygnets during their early life. Almost all the birds are banded and of known age, and reproduction is followed in detail. Laying dates did not change over time, but clutch size showed an average increase of 0.078 eggs per swan generation, which equals about 7.2 years. This change in clutch size is consistent with evolutionary theory, which predicts that with relaxation of constraints on food and predation risk, swans with larger clutch sizes should be favored by natural selection.

In a few cases, species of waterfowl show likely evolutionary responses in morphology to climatic change. Between 1971 and 1996 in northern Germany, for example, the Common Goldeneye showed increases in body mass, clutch size, and hatching success but a decrease in wing length. It is possible that shortened wing length reflects selection resulting from reduction in migration distance between breeding and wintering areas.

The Mallard is notorious for hybridization with close relatives such as the Mottled and Black Ducks in North America and many other congeners worldwide. In most cases, the increased frequency of hybridization seems to be the result of human modification of landscapes and introductions of Mallards to new regions. Climatic warming, however, appears likely to have led to extensive hybridization between the Spot-billed Duck and the Mallard in eastern Russia. The Spot-billed Duck has spread northward about 500 kilometers, bringing it into broad overlap with the Mallard.

## Storks and Cranes

A number of experiments with White Storks in Europe, beginning in the 1930s, have attempted to determine whether migration patterns are hereditary. White Storks normally migrate diurnally and in large flocks. Several studies have indicated that juvenile birds held in aviaries until all adults have departed tend to follow southward migration routes that in some cases are similar to those of adults and in other cases deviate significantly in a more westerly path. A more ambitious set of experiments on White Stork migration used juvenile birds equipped with satellite transmitters to permit detailed tracking of their movements. In 2000, four juvenile birds were equipped with transmitters in nests near Kaliningrad, Russia, and they subsequently initiated southward migration with adults. One apparently died soon after fledging, but three followed the traditional route used by adults around the western end of the Black Sea, across Turkey, and around the eastern coast of the Mediterranean Sea to wintering areas in East Africa. Other juvenile storks carrying transmitters were released after adults had departed or at localities roughly 30° to 52° east of Kaliningrad, where White Storks do not breed, but at the proper time for southward migration. These birds, released in 2001 and 2002, all tended to move south, but directions and distances of their movements varied considerably, and none reached wintering areas in East Africa. These observations, together with those of earlier studies, were most consistent with the idea that White Storks have only an approximate inherited migration program and that the precise migration pattern achieved by the species is dependent on social interaction and learning.

White Storks have begun to overwinter across much of southern Europe since the 1980s. The wintering areas for these birds in sub-Saharan Africa have become less favorable because of drought, wetland degradation, and hunting by humans. The change in migratory pattern might thus be partly the result of natural selection. In similar fashion, the Common Crane, formerly wintering from southern Spain and Turkey southward, has recently established a wintering population in northern France.

## Herons and Egrets

The establishment, spread, and development of migration by the Cattle Egret in the New World presents interesting questions about the origin of migration patterns. This species apparently became established in northeastern South America by natural dispersal from Africa in the late 1800s. At first a permanent resident of tropical livestock ranches, it appeared in North America in the early 1940s and began breeding in Florida in the early 1950s. Since then its breeding population has spread throughout much of the United States and even southern Canada, and a strong pattern of seasonal migration has become established between northern breeding areas and more-southern areas of wintering and permanent residency. The species shows extensive postbreeding dispersal, and appearance of regular migration probably indicates that basic genetic tendencies for migration were present when the species entered North America and that as migratory behavior began to appear, selection refined it into its present pattern. Northward spread of migratory Cattle Egrets is still occurring.

The Little Egret of the Old World is behaving much like the Cattle Egret. It has expanded its range northward in Europe and now breeds in England and Ireland. Although many European birds migrate south to winter in sub-Saharan Africa and coastal Mediterranean areas, some nonmigratory birds now winter in southern Britain. Many continental birds also apparently migrate northwest to the British Isles to winter. The Little Egret has also established a New World breeding colony on Barbados, in the West Indies.

## Seabirds

In 1995–1996, Leach's Storm-Petrel was discovered breeding on Dyer Island, off the southern coast of South Africa. Nesting was previously known only in North Atlantic and North Pacific localities. How recently these colonies were first established is uncertain, and whether this population is genetically distinct from those in the Northern Hemisphere is also uncertain. The population of this species on Guadalupe Island, Mexico, however, shows seasonally segregated subpopulations that likely are genetically distinct, as are several such populations of the widespread Band-rumped Storm-Petrel.

## Summary

Waterbirds, even some long-lived species, are also showing evolutionary changes that appear to be responses to climate change. European Great Crested Grebes show patterns of change in migratory schedules and wintering areas that appear to be genetically based. Shorebirds and waterfowl exhibit patterns of genetic variability that may influence their response to climatic change. A few species of ducks show morphological change or patterns of hybridization that appear to result from climatic change. Changes in wintering patterns of several storks and egrets are likely to reflect evolutionary influences. An evolutionary shift might be involved in the establishment of Leach's Storm-Petrel, a widespread seabird of the North Atlantic and North Pacific, as a breeding bird on an island off the South African coast.

## KEY REFERENCES

Adriaensen, F., P. Ulenaers, and A. A. Dhondt. 1993. "Ringing recoveries and the increase in numbers of European Great Crested Grebes *Podiceps cristatus*." *Ardea* 81:59–70.

Alisauskas, R. T. 1998. "Winter range expansion and relationships between landscape and morphometrics of midcontinent Lesser Snow Geese." *The Auk* 115:851–862.

Battley, P. F. 2006. "Consistent annual schedules in a migratory shorebird." *Biology Letters* 2:517–520.

Bêty, J., J.-F. Giroux, and G. Gauthier. 2004. "Individual variation in timing of migration: Causes and reproductive consequences in greater snow geese (*Anser caerulescens atlanticus*)." *Behavioral Ecology and Sociobiology* 57:1–8.

Both, C., T. Piersma, and S. P. Roodbergen. 2005. "Climatic change explains much of the 20th century advance in laying date of Northern Lapwing *Venellus vanellus* in The Netherlands." *Ardea* 93:79–88.

Charmantier, A., C. Perrins, R. H. McCleery, and B. C. Sheldon. 2006. "Evolutionary response to selection on clutch size in a long-term study of the mute swan." *The American Naturalist* 167(3):453–465.

Chernetsov, N., P. Berthold, and U. Querner. 2004. "Migratory orientation of first-year white storks (*Ciconia ciconia*): Inherited information and social interactions." *Journal of Experimental Biology* 207:937–943.

Gordo, O. and J. J. Sanz. 2006. "Climate change and bird phenology: A long-term study in the Iberian Peninsula." *Global Change Biology* 12:1993–2004.

Gunnarsson, T. G., J. A. Gill, T. Sigurdjörnsson, and W. J. Sutherland. 2004. "Arrival synchrony in migratory birds." *Nature* 431:646.

Kulikova, I. V., Y. N. Zhuravlev, and K. G. McCracken. 2004. "Asymmetric hybridization and sex-based gene flow between eastern spot-billed ducks (*Anas zonorhyncha*) and mallards (*A. platyrhynchos*) in the Russian far east." *The Auk* 121:930–949.

Ludwichowski, I. 1997. "Long-term changes of wing-length, body mass and breeding parameters in first-time breeding females of goldeneyes (*Bucephala clangula clangula*) in northern Germany." *Vogelwarte* 39:103–116.

Pulido, F., P. Berthold, and A. J. van Noordwijk. 1996. "Frequency of migrants and migratory activity are genetically correlated in a bird population: Evolutionary implications." *Proceedings of the National Academy of Sciences USA* 93:14642–14647.

# PART V

## Prospects

In this concluding pair of chapters, we first consider whether or not the adaptive capabilities of migratory birds are adequate to meet the challenges of global climate change. We conclude with a survey of national and international efforts that are directed at protection of migratory birds in the face of environmental change.

# Chapter 19

## *Capacity for Adjustment by Migratory Birds*

Little doubt remains that migratory birds of many sorts are responding to global climate change. We now consider the degree to which ecological and evolutionary responses as we now understand them are sufficient to ensure the survival of migratory birds in the face of the global changes that are certain to occur. The ability to adapt to change depends on the degree of plasticity of behavioral and physiological systems to make adjustments to short-term changes and the ability of populations to evolve in the face of changing selective pressures. For migratory birds, evolutionary responses are influenced by selective pressures in different geographical regions, in some cases widely separated, at different stages of the annual cycle.

The potential for adaptive change in migration schedules and routes depends in large measure on the length of the migration path and the geographical and climatic challenges it presents to migrating birds. For long-distance migrants in particular, the complexities of climatic change may lead to opposing selective advantages at different times and locations in the annual cycle. In addition, selection across a given population may be influenced by the extent to which its members follow a common migratory route from a specific breeding area to a specific nonbreeding range.

## Life History Characteristics and Response Capacity

Basic life history characteristics constrain the ability of migratory birds to respond to climatic change. Migratory birds vary greatly in body size, longevity, potential population growth rate, and habitat specificity. These factors all influence the risk of extinction by birds faced with rapid environmental change. An analysis of more than 1000 bird species considered to be at risk of extinction suggests that habitat loss alone is the greatest threat for about 43 percent of species, especially those with specialized habitat requirements. Many of these species are also small bodied and have short generation times. Many are also permanent residents, altitudinal migrants, or short-distance migrants of tropical regions. A few long-distance migrants, such as Kirtland's Warbler in North America, also fall into this category. Although this analysis did not directly address the risk due to climate change, it does suggest that regardless of other life history features, habitat specialists face the most serious threat from changing climate.

Large-bodied species with long generation times and low population growth rates, such as cranes and large seabirds, face the most severe difficulties in evolutionary response to rapid environmental change. They are also most likely to experience direct persecution by humans or predators. The rate of adaptation to climatic change appears to be set primarily by the potential for microevolution, the rate of which is apparently low in vertebrates, and especially so in long-lived species. Even if genetic variability is available in the population and if selection for some genotypes is strong, rapid climate change may lead to low reproductive success and high mortality. Thus, the risk for many of these species is that populations may fall below the viability threshold, leading to a decline to extinction.

## Migratory Connectivity and Response Capacity

Migratory birds vary greatly in the degree to which individuals from a specific breeding area spend the nonbreeding period in a specific nonbreeding area. Migratory connectivity refers to this pattern. Species with high connectivity are presumed to have rigidly defined migration patterns, with genetic variation in their choice of routes being low. Such species are likely to respond only very slowly to climatic changes that affect the geographical distribution of breeding and nonbreeding habitats. Species with low connectivity, on the other hand, presumably have greater variability in the genetic control of migration patterns and perhaps greater phenotypic plasticity, both of which suggest that they can respond more quickly to climatic change. Knowledge of these patterns of connectivity also should be useful in addressing the threats to survival posed by climate change that affects breeding and wintering habitats.

Among North American birds, many long-distance migrants, such as parulid warblers, exhibit moderate to high connectivity. Stable isotope signatures from feathers collected from birds in different parts of the winter range indicate that American Redstarts show moderately strong connectivity of wintering and breeding locations. Birds from the Pacific Northwest, for example, tended to winter in western Mexico, those from the upper Midwest in Central America, those from the Northeast in the West Indies, and those from the Southeast in Trinidad and Tobago. Eastern birds thus showed a latitudinal separation in winter somewhat similar to that shown in the breeding range. For the Yellow Warbler, one of the most widely distributed long-range migrants in North America, connectivity was not as strong, but strong segregation exists between eastern and western populations that use roughly parallel north–south migration routes. Black-throated Blue Warblers show moderate connectivity, with birds breeding in northeastern North America tending to winter in Jamaica and Cuba, and with those breeding in more-southern areas wintering in Hispaniola and Puerto Rico.

Long-distance migrant shorebirds and other waterbirds, particularly those that tend to follow continental coastlines, tend to show high connectivity. The Dunlin, a long-distance migrant that breeds in the Arctic and winters from subtropical areas of the Northern Hemisphere to the temperate Southern Hemisphere, shows moderate to strong connectivity. In North America, the Whooping Crane exhibits very strong connectivity involving breeding and wintering ranges, as do Sandhill Cranes.

Short-distance migrants show varying degrees of connectivity. Willets breeding in the western Great Basin tend to migrate to coastal sites and estuaries near San Francisco Bay. They show little tendency to move once they reach the coast, and they tend to be faithful to specific wintering areas from year to year. The winter range of this population differs from that of other Willets, suggesting that various subpopulations of this species may show high connectivity of breeding and wintering areas. The White-throated Sparrow, a short-distance migrant, shows almost no connectivity, with birds from many breeding areas mixing together in wintering areas. In contrast, stable isotope analysis shows a high degree of connectivity of breeding and wintering habitats in the Coastal Plain Swamp Sparrow. This subspecies breeds in coastal marsh areas from Virginia to southern New York and winters largely in coastal North Carolina marshlands.

## Phenotypic Plasticity and Ecological Adjustment

Phenotypic plasticity is the basis for immediate ecological adjustment to changing climatic conditions. In North America, four times as many short-distance migrants as long-distance migrants show spring arrival patterns correlated with temperatures, suggesting that most advances in arrival represent ecological adjustments. Nevertheless, a few long-distance migrants showed correlations with spring temperatures in their arrival patterns, and nearly three-quarters showed influences of the North Atlantic Oscillation (NAO), tending to arrive earlier during positive phases of this climatic cycle. The influence of the positive NAO seems to relate to conditions that favor earlier departure from wintering areas or speedier passage along southern portions of the migration route. Thus, even long-distance migrants show some plasticity in responses to weather patterns.

Nevertheless, other studies suggest that responses of long-distance migrants to climate change are weaker than those of short-distance migrants. As we have seen, long-distance migrants in North America and Europe have not adjusted their migration schedules to the degree shown by short-distance migrants. Whether this failure is due to lack of behavioral plasticity or genetic variability, or both, is uncertain. In Europe, the long-distance migrant Garden Warbler and the short- to medium-distance migrant Blackcap show no difference in degree of additive genetic variation in timing of fall migration. This observation suggests that the slower adaptive responses of long-distance migrants to climate change reflect a lower degree of phenotypic plasticity, rather than lack of genetic variability.

Phenotypic plasticity is high in migratory waterfowl. At Delta Marsh in Manitoba, Canada, for example, Canada Geese, Tundra Swans, and numerous species of ducks showed significantly earlier spring arrivals in years with warmer spring temperatures. In Europe, different environmental factors control the departure of Pink-footed Geese from their winter home in Denmark and from stopover locations in Norway. It appears that both day length and temperature conditions determine the time of departure from Denmark, but temperatures thereafter determine migration scheduling. Phenotypic plasticity in this species appears adequate to enable the species to respond to climatic change.

Some birds may possess greater potential for ecological adjustment than is apparent. For example, a resident subspecies of the African Stonechat shows migratory restlessness behavior at times, even when held under a constant photoperiod. Although the specific population was nonmigratory, other African populations of the species show altitudinal movements, and this behavior may reflect adaptation for dispersal or movement of birds in response to drought or other unusual weather conditions. Thus, some genetically based mechanisms related to migration behavior may be flexible enough to allow for ecological response to changing weather patterns.

## Capacity for Evolutionary Adjustment

Looking at migratory capability in a phylogenetic context, it appears that migratory capability has evolved repeatedly. There is little evidence for a coadapted complex of migratory traits basic to all higher birds. Migration as a response to seasonal climatic conditions is likely to have appeared in the earliest birds, or even their more-mobile ancestors. Some mechanisms of diurnal or nocturnal orientation might have existed in early birds, but it is more likely that complex mechanisms of physiological preparation for migration and behavioral orientation during migration have evolved independently in different lineages. It is also likely that physiological and behavioral capabilities for migration were lost in some lineages.

Families of modern birds differ considerably in the degree and type of migratory ability. Most cisticolas (Cisticolidae), a family of 111 species widely distributed in Africa, Asia, and Australasia, are permanent residents. Short-distance migration is shown by only a few species, in spite of the fact that many occur in areas of strongly seasonal climate. In other families, migration occurs in many species but tends to be mostly nocturnal in some, such as the Old World warblers (Sylviidae), and diurnal in others, such as the pipits and wagtails (Motacillidae). This suggests that all modern birds do not have equal genetic potentials for adjusting migration patterns.

Some capacity for evolutionary adjustment of migration patterns exists, however, in many short-distance and long-distance migrants. Some avian physiologists consider this capacity to be considerable. For some species, as Peter Berthold put it, "Birds are . . . evidently capable of adapting to changing environmental conditions by shifting from almost full sedentariness to exclusive migratoriness and vice versa based on selection." Thomas Alerstam even regarded long-distance migrants as having high intrinsic ability to adjust migration patterns through evolution: "Given this great flexibility one can expect that migration to a large degree evolves according to the ecological opportunities."

Within many groups of closely related species, shifts between permanent residence and short- or long-distance migration are frequent, and populations that are partially migrant appear to be able to switch fairly rapidly to fully resident or fully migrant. Within some species, such as the Blackcap, there are regional populations that are resident, partially migrant, and short-distance and long-distance fully migrant. In this species, experimental hybridization of individuals from sedentary and fully migrant populations has shown that the offspring show intermediate patterns of migratory behavior. Experimental selection over four to six generations can also transform partial migrant lineages of Blackcaps into fully resident and fully migrant lineages. In these birds, migratory behavior has a genetic basis with a threshold across which selection can quickly change the pattern seen in a population.

Other aspects of migratory behavior and physiology also appear to be under strong genetic control. The timing and directional orientation of migration appear to be under strong genetic control in some passerines. Hybridization experiments with the Common Redstart and Black Redstart of Europe, which have different migration schedules, result in offspring with intermediate patterns of migration timing. An elegant series of experiments using the Blackcap showed that hybrid offspring of birds migrating in different directions showed intermediate directional choice. Furthermore, the duration of migratory disposition in experimental birds correlates with the length of the migration route followed by birds in nature. Thus, these aspects of migratory behavior and physiology seem to be under genetic control.

How rapidly natural selection can adjust genetically based migration patterns in the face of climatic change, however, certainly varies among migratory species. Short-distance and partial migrants should be able to adapt quickly by shifts to partial migration and permanent residency, respectively, as climates warm. Whether long-distance migrants can keep up with the rate of climate change is less certain. Nevertheless, many long-distance migrants may have genetic variability on which selection can act. As we noted earlier, the Garden Warbler, a long-distance migrant, shows as much additive genetic variance as the short-distance migrant Blackcap.

For bird species breeding in polar environments, many of which are long-distance migrants, the ability to adjust physiology and behavior to changing seasonal conditions on the breeding range is of key importance. The timing of arrival and nesting must be matched to phenological events so that the abundance of food during the nestling and fledgling periods is maximized. Birds that fail to adjust to the phenology of food resource availability are likely to experience reduced reproductive success.

## Overall Response to Climatic Change

Possible changes in bird community composition across Europe were explored by predictive climatic modeling, using the HadCM3 general circulation model (GCM). In this exercise, some 324 land bird species were used. This model considered for each locality the proportion of species that were migratory in some portion of their range (migratory propensity), the proportion of species capable of migration that actually did migrate (migratory activity), and the proportion of actually migrating species relative to all birds (migratory proportion). Using climate data from AD 1961–1990 as a baseline, they projected climate change to AD 2051–2080 with the GCM model. The projections also included changes in precipitation patterns. In these projections, the bird community could change by adding or losing species with different migratory propensities and by changes in migratory activity for species that showed facultative migration capability.

Modeling results showed generally that warming of winter temperatures led to little change in the proportion of species that could migrate but led to large decreases in the fraction of the bird community that actually did migrate. Warming spring conditions, on the other hand, favored increased migratory activity, both by an increase in species capable of migration and an increase in migratory activity by those that could. These responses tended to be stronger or weaker in different parts of Europe. In southern and southwestern Europe, as the climate became more strongly seasonal, the fraction of the bird community that was migratory tended to increase, primarily because of gain of potentially migratory species. In northern Europe, with warming, migration tended to decrease, as facultatively migratory species switched to permanent residency.

Overall, however, European bird communities have not changed as fast as their climate has changed. A recent analysis attempted to determine the overall response of breeding birds in France to changing climate. For each of 105 species, investigators calculated a species temperature index (STI), which was determined as the mean temperature experienced by the species over its European breeding range from March through August. From these values, they determined a community temperature index (CTI) as the mean of STIs for species breeding at 1514 locations surveyed in the

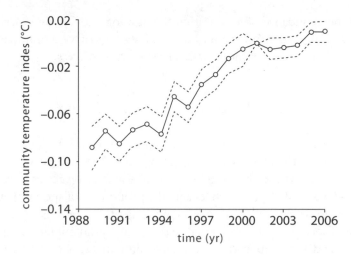

FIGURE 19.1. Average community temperature index for 105 species of birds at locations surveyed in the French Breeding Bird Survey from 1989 to 2006. (From Devictor, V., R. Julliard, D. Couvet, and F. Jiguet. 2008. "Birds are tracking climate warming, but not fast enough." *Proceedings of the Royal Society of London B* 275:2745.)

French Breeding Bird Survey. Over the period 1989–2006, the researchers found that the CTI increased in a generally steady manner (Figure 19.1). The increase did not occur at the same rate as the increase in site temperature, however (Figure 19.2). Over the 18-year period, bird communities realized only 54 percent of the expected change, corresponding to a 97-kilometer lag in northward shift.

## Summary

Long-distance migrants tend to show moderate to strong connectivity between breeding and wintering areas, while short-distance migrants are quite variable. Many short-distance migrants, both land birds and some species of waterbirds, show high phenotypic plasticity in timing of migration. Most long-distance migrant songbirds exhibit lower phenotypic plasticity. Although migration is shown in the majority of higher bird taxa, little evidence exists for a coadapted complex of genetic traits basic to all higher birds. Many groups of closely related species exhibit patterns ranging from permanent residency to long-distance migration and show a potential for rapid transition among these patterns. Nevertheless, long-distance migrants ap-

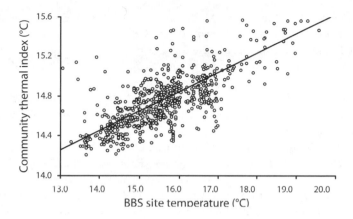

FIGURE 19.2. Correlation between the average community temperature index and site temperature for 722 locations in France surveyed in 2005. (From Devictor, V., R. Julliard, D. Couvet, and F. Jiguet. 2007. "French birds lag behind climate warming." *Nature Precedings*:hdl:101`01/npre.2007.1275.1.)

pear to be at greater risk from changing climate. Analysis of bird community responses to climate change suggest that adjustments involving changes in migration tendency within and between species were incomplete. Evidence also exists that some long-distance migrants are failing to adjust the timing of migrations to changing climate.

## KEY REFERENCES

Alerstam, T., A. Hedenstron, and Susanne Äkesson. 2003. "Long-distance migration: Evolution and determinants." *Oikos* 103:247–260.

Berteaux, D., D. Réale, A. G. McAdam, and S. Boutin. 2004. "Keeping pace with fast climate change: Can arctic life count on evolution?" *Integrative and Comparative Biology* 44:140–151.

Berthold, P. 1998. "Bird migration: Genetic programs with high adaptability." *Zoology* 101:235–245.

Boulet, M. and D. R. Norris. 2006. "The past and present of migratory connectivity." *Ornithological Monographs* 61:1–13.

Devictor, V., R. Julliard, D. Couvet, and F. Jiguet. 2008. "Birds are tracking climate warming, but not fast enough." *Proceedings of the Royal Society of London B* 275:2743–2748.

Helm, B. and E. Gwinner. 2006. "Migratory restlessness in an equatorial nonmigratory bird." *PLoS Biology* (online) 4:0611–0614.

Lemoine, N. and K. Böhning-Gaese. 2003. "Potential impact of global change on species richness of long-distance migrants." *Conservation Biology* 17:577–586.

Norris, D. R., M. P. Wunder, and M. Boulet. 2006. "Perspectives on migratory connectivity." *Ornithological Monographs* 61:79–88.

Pulido, F. and M. Widmer. 2005. "Are long-distance migrants constrained in their evolutionary response to environmental change?" *Annals of the New York Academy of Sciences* 1046:228–241.

Schaefer, H.-C., W. Jetz, and K. Bohning-Gaese. 2008. "Impact of climate change on migratory birds: Community reassembly versus adaptation." *Global Ecology and Biogeography* 17:38–49.

Visser, M. E. 2008. "Keeping up with a warming world: Assessing the rate of adaptation to climate change." *Proceedings of the Royal Society B* 275:649–659.

Webster, M. S. and P. P. Marra. 2005. The importance of understanding migratory connectivity and seasonal interactions. Pp. 199–209 in R. Greenberg and P. P. Marra (Eds.), *Birds of Two Worlds*. Johns Hopkins University Press, Baltimore.

# Chapter 20

# *Conservation in an Era of Global Change*

Protecting migratory birds in an era of changing global climate will require greater understanding of the changes that are likely to occur in breeding and wintering areas, as well as in the networks of stopover areas used in migration. The structure and dynamics of virtually all of the earth's ecosystems will be altered to some degree. The development of protected areas will be difficult because changing climates will tend to force birds from areas presently under protection to unprotected areas that have become more favorable because of altered conditions.

Key strategies for conservation will be (1) monitoring the ecological and evolutionary responses of migratory birds to climate change and (2) devising international plans for habitat protection that incorporate the flexibility necessary to keep pace with environmental change.

## Status of Migratory Species

The changing world environment has accelerated the endangerment and extinction of species. According to the 2008 assessment by the International Union for Conservation of Nature (IUCN), 134 species of birds, or about 1.3 percent, are known to have become extinct since AD 1500. The IUCN's Red List ranks species of conservation concern in four categories: critically endangered, endangered, vulnerable, and near threatened.

249

TABLE 20.1. Endangered migratory birds on the 2008 IUCN Red List.

| | Total Species | Migratory Birds[1] | Land and Freshwater Birds[1] | | | Oceanic Birds |
| --- | --- | --- | --- | --- | --- | --- |
| | | | Americas | Eurasia/ Africa | Asia/ Pacific | |
| Critically endangered | 190 | 29 | 3 | 1 | 8 | 17 |
| Endangered | 361 | 52 | 5 | 8 | 17 | 24 |
| Vulnerable | 671 | 117 | 29 | 17 | 41 | 37 |
| Near threatened | 835 | 113 | 39 | 32 | 25 | 25 |
| Total | 2057 | 311 | 76 | 58 | 91 | 103 |

[1]Some migratory species occur in more than one land region.

Critically endangered species are considered to have a 50 percent chance of extinction in 10 years or three generations. Endangered species are estimated to have a 20 percent chance of extinction in 20 years or five generations. Vulnerable species are considered to possess a 10 percent chance of extinction in 100 years. Near threatened species are those that are close to qualifying for vulnerable or more endangered status or are likely to qualify in the near future. About 1222 species (12.3 percent) fall into the first three categories, and 835 (8.4 percent) fall into the near threatened category (Table 20.1). Some 190 species (1.9 percent) are considered to be critically endangered, with the likelihood that about 15 of these are actually extinct.

The IUCN Red List includes 311 migratory species in all four categories; 208 are migratory land or freshwater birds, and 103 are seabirds (Table 20.1). Land and freshwater birds listed as being at risk correspond to about 8.0 percent of all migratory species. The greatest numbers of migratory species at risk, 91, are in the Asia–Pacific region; the fewest, 58, are in western Eurasia and Africa. For the world's 234 seabirds (see Chapter 1), however, about 44 percent of migratory species are classified as at risk, with about a third falling in the top three categories of endangerment. Overall, based on the numbers of bird species noted in Chapter 1, only about 12.0 percent of all migratory species are considered to be at significant risk of extinction, compared with about 23.8 percent of nonmigrants. Since 1988, 33 Red List species have shifted to a more endangered status, whereas only 6 have improved.

Of birds that have become extinct since about AD 1600, only a handful, such as the Passenger Pigeon and Carolina Parakeet, were migratory birds of continental areas. Among land birds of North America, the IUCN listing includes only the Black-capped Vireo and Golden-cheeked Warbler as endangered. The National Audubon Society considers two additional migratory North American land and freshwater birds to be endangered: the Piping Plover and Kirtland's Warbler. None of the North American species listed by these groups appears to have declined because of recent climate change.

## Birds at Highest Risk from Changing Climate

Many endangered birds are resident species that occur in localized areas, including oceanic islands, freshwater lakes, and isolated mountain systems. Others are ground-dwelling birds of forest areas, particularly in the tropics, that are averse to crossing areas of nonforest habitat. For these species, habitat fragmentation is a serious challenge because of the heightened risk of extinction of local populations due to stochastic processes. Climatic change is an added challenge to their survival. The numbers of these residents at risk of extinction are substantial, and the efforts required for their protection will be enormous. Some biologists estimate that over a quarter of sedentary bird species are at risk of extinction.

Migratory birds, however, present unusual conservation challenges because of their dependence on a range of seasonal environments. Our review of migratory birds in various world environments reveals that certain groups are especially likely to experience severe population declines due to changing climate.

Long-distance migrants that depend on breeding areas that are experiencing climatic change that alters the seasonal optimum of resources for rearing young are at high risk. For these species, conditions of the nonbreeding range may not provide cues for altering migration schedules to maintain synchrony of arrival and nesting with the optimal resources for reproduction. Among these birds, long-distance migrants between western Eurasia and sub-Saharan Africa appear to be at greatest risk, largely due to the substantial climate change that has occurred in their breeding ranges. In western Eurasia, in general, warming climate is projected to shift breeding ranges farther north while modifying nonbreeding ranges very little. The consequence is thus a lengthening of the migration route and increase in the energy demand for migration. Changes in atmospheric conditions due to climate change may also challenge the energetic capabilities of migrants that must cross major ecological barriers such as the Sahara. We have seen that severe declines are occurring in the populations of many of these species. This relationship probably exists to some degree for long-distance migrants in eastern Asia and North America. Substantial declines have also occurred in Neotropical long-distance migrants, particularly those breeding in eastern North America. North American Breeding Bird Survey data for 1980–2005 show that 62 percent of Nearctic–Neotropical migrants there have declined.

Long-distance migrant shorebirds face similar, and perhaps even greater, challenges. The high northern latitudes where most breed have experienced some of the greatest climatic changes, yet conditions in their nonbreeding ranges give few cues that could enable them to adjust migration schedules. Furthermore, wetlands that serve as staging and stopover areas are also being affected by climate change and sea level rise. As we saw in Chapter 12, populations of about half of all shorebird species are declining.

Long-distance migrants depend on staging and stopover sites that permit birds to rest and restore energy supplies for continued migration. Many of these are also being affected by climatic change. For land birds, key staging and stopover areas tend to be areas of productive natural vegetation bordering oceans, lakes, or deserts. For shorebirds, coastal and interior marsh and shore habitats are essential to migration. Efforts are needed to protect networks of such sites that can provide food and shelter even as climates change.

Breeding habitats of some migratory birds are likely to be severely reduced or eliminated by climatic change. Tropical cloud forest birds, many of which show altitudinal movements, are likely to lose breeding habitat. Many of these species are frugivores or nectarivores and depend on the seasonal progression of resources over an altitudinal gradient. Birds of the alpine tundra zone in temperate regions are also likely to lose breeding habitat. Marsh birds and shorebirds, already suffering from habitat loss due to human activities, will lose habitat in areas of warming and drying climate.

Migratory and nomadic species that depend on a certain frequency of recurrent favorable conditions within a region are at high risk due to climatic change leading to warmer and drier conditions. Nomadic and semi-nomadic birds in areas of Africa, southern Asia, and Australia are already beginning to be affected.

Birds breeding on oceanic islands, both seabirds and land birds, are the most numerous species listed as endangered. For many of the land birds, threats to survival are most often due to human alteration of natural habitats and the introduction of predators, competitors, and diseases. For seabirds, however, changes in ocean currents, upwelling systems, and tem-

perature profiles due to altered climatic conditions are reducing food supplies in many locations, leading to reproductive failures. How easily many
of these seabirds can shift colony sites to more favorable areas is uncertain.

## Monitoring the Ecological Status of Migratory Birds

Breeding bird populations, both migratory and nonmigratory, are now being monitored by extensive surveys in North America and Europe. In the
early 1960s, scientists at the Migratory Bird Population Station in Laurel,
Maryland, conceived a roadside survey program for North American breeding birds. The North American Breeding Bird Survey (BBS) was tested in
1965 and initiated in 1966 with about 600 surveys in the eastern United
States and Canada. The survey spread to the Great Plains states and Canadian Prairie Provinces in 1967. By 1968, about 1000 routes were being surveyed across southern Canada and the United States. In the 1980s, the BBS
was extended to Alaska and the Yukon and Northwest Territories of Canada. Now, more than 2900 BBS routes are surveyed annually in the continental USA and Canada.

BBS surveys involve 3-minute point counts at fifty stops along each
route. At each stop, the observer records all birds heard or seen within a ra-
dius of 0.40 kilometer. Surveys thus produce indices of relative abundance
rather than complete censuses of breeding birds. BBS data are analyzed to
produce relative abundance and population trend maps for species that are
adequately sampled by sight and sound.

The Pan-European Common Bird Monitoring project, initiated in
2002, is intended to monitor breeding populations of common birds as indicators of general environmental health across Europe. The project is undertaken jointly by BirdLife International and the European Bird Census
Council. BirdLife International is a worldwide partnership of nongovern-
mental conservation organizations in more than 100 countries. Its goal is
to promote conservation of biodiversity and sustainable use of natural resources through a focus on birds and their habitats. The European Bird
Census Council consists of European ornithologists who represent national organizations that monitor bird populations and distributions. Its
goal is to encourage activities such as the development of bird atlases and
the monitoring of common birds as indicators of the ability of European
landscapes to support wildlife at large.

About thirty-six European countries are participants in the Pan-
European Common Bird Monitoring project, which plans to coordinate

more than twenty national and regional breeding bird surveys across Europe and to encourage the establishment of new national monitoring schemes. The objective is to use the data from these surveys to derive general indices and indicators of environmental health.

## Management of Migratory Bird Habitat

Major efforts are being undertaken throughout the world to achieve the protection of migratory birds and their habitats. BirdLife International is also developing a network of Important Bird Areas (IBAs), sites identified as critically important for conservation of bird populations. Several categories of species are considered in identifying such sites. First, sites that hold populations of species listed on the IUCN Red List as critically endangered or endangered, or significant populations of listed vulnerable species, qualify as IBAs. Second, sites that contain significant numbers of species with very restricted ranges qualify. Third, sites that ensure representation of species restricted to a regional ecosystem type are included. Finally, and of special significance, the network includes sites that support large breeding or nonbreeding congregations of waterbirds and sites that constitute bottlenecks heavily used by species during migration. This effort identified more than 4000 sites used by more than 700 migratory bird species. Only a few of these sites have formal protection, so one objective of the program is to increase the extent of protection.

The Convention on the Conservation of Migratory Species of Wild Animals was developed in Bonn, Germany, in 1979 and became effective in 1985. Commonly known as the Bonn Convention or Convention on Migratory Species, it requires member states to conserve migratory species of all animal groups and their habitats. The convention lists seventy-two endangered migratory birds and several hundred species of other birds, especially waterbirds, that would benefit from international cooperation. This convention was promoted by the United Nations Environment Programme. Since 1985, the convention has been ratified by more than a hundred countries throughout the world. Under the convention, treaties and memoranda of understanding are arranged to further conservation of the species covered.

The Convention on Wetlands, known commonly as the Ramsar Convention, was signed in Ramsar, Iran, in 1971. This treaty provides for international conservation of wetlands and their species. In 2008 Ramsar had 158 member states and 1752 sites designated as Wetlands of International

Importance. The number of designated sites varies enormously, with the United Kingdom (166), Mexico (86), and Australia (64) having the most. Originally focused on wetlands as habitats for waterbirds, the current program considers broader aspects of conservation and wise use.

## Arctic Regions

In 1996 the Arctic Council was formed to provide a forum for cooperation, coordination, and interaction among Arctic states, including indigenous peoples, on issues of sustainable development and environmental protection. The member states are Canada, Denmark (representing Greenland and the Faeroe Islands), Finland, Iceland, Norway, the Russian Federation, Sweden, and the United States of America. Other participants include councils of indigenous peoples, including Aleut, Inuit, Arctic Athabascan, Gwich'in, Saami, and Russian Arctic peoples. The biodiversity working group of the Arctic Council is Conservation of Arctic Flora and Fauna (CAFF). Its mission is to address the conservation of Arctic biodiversity and communicate its findings to the governments and residents of the Arctic, helping to promote practices that ensure the sustainability of the Arctic's living resources.

One of CAFF's expert groups is the Circumpolar Seabird Group, which provides a forum to facilitate conservation, management, and research activities relating to seabirds. This group sponsored a workshop on conservation of migratory Arctic birds in 2000. Their report specifically identified nine species of globally threatened migratory birds: Lesser White-fronted Goose, Red-breasted Goose, Baikal Teal, Steller's Eider, Steller's Sea-Eagle, Siberian Crane, Eskimo Curlew, Bristle-thighed Curlew, and Spoon-billed Sandpiper. The most recent meeting of the Circumpolar Seabird Group was held in Greenland in February 2008, and various issues, including population declines of Glaucous Gulls and the relation of murre populations to climate change, were considered.

## North and South America

Most migrating birds found in Canada and the United States are protected under the Migratory Birds Convention of 1916 between the two countries, which was followed by subsequent implementing legislation. At first, Migratory Bird Protection Acts excluded some birds considered pests, including pelicans, cormorants, hawks, and owls. Several amendments to American and Canadian Migratory Bird Protection Acts have since given protection to these species and have also permitted traditional wildlife harvests by aboriginal peoples of the United States and Canada. In the United States, the Migratory Bird Protection Act also has amendments to reflect more-recent conventions with Mexico, Japan, and Russia, which cover particular sets of species.

Partners in Flight is a cooperative effort between governmental agencies and private organizations to work toward protection of land birds throughout the Americas. Initially, the plan focuses on 448 native land birds, both migratory and resident, of the United States and Canada, with full participation of Mexico expected to add 450 additional species. This plan identifies 100 "watch list" species at risk due to habitat loss or other challenges and 28 species of critical concern. Population estimates are presented for these species, and management goals are indicated.

The North American Waterbird Conservation Plan, entitled *Waterbird Conservation for the Americas*, covers North America from Arctic Canada and Alaska to Panama and the Caribbean, as well as Pacific islands under United States jurisdiction. This plan is under continuous development, beginning with colonial seabirds and freshwater birds but eventually to extend to 210 species of birds occurring in twenty-nine countries throughout the region. The plan addresses threat to waterbirds from all sources: habitat destruction, alien predators and invaders, pollution, mortality from fisheries activities, and influences of other birds. In cooperation with BirdLife International, the plan is also being extended to South America under support of the United States Neotropical Migratory Bird Conservation Act passed in 2000. The Canadian participation in this effort is described in *Wings over Water: Canada's Waterbird Conservation Plan*. The plan covers 93 species of marine, freshwater, and marsh birds, about 30 percent of which are experiencing population declines.

The North American Waterfowl Management Plan was created in 1986 by American and Canadian governments to encourage waterfowl populations by protection, restoration, and enhancement of waterfowl habitats. Waterfowl populations are notoriously variable from year to year, depending heavily on weather and wetland conditions during the breeding season. Total duck populations in 2005 were estimated to be 31.7 million, well above the numbers in the late 1980s that led to the development of the management plan, but almost 10 million below the high population estimated in 2000. Populations of most species of geese, but especially midcontinent Snow Geese, have increased in recent decades. Tundra Swans have generally increased since the 1970s.

The Canadian Shorebird Conservation Plan focuses on forty-seven species of shorebirds that breed in Canada. These include thirty-three sandpipers and three phalaropes (Scolopacidae), seven plovers (Charadriidae), two oystercatchers (Haematopodidae), the American Avocet, and the Black-necked Stilt. Of thirty-five of these species for which good population trend data are available, twenty-eight show population declines, emphasizing the critical need for active conservation efforts. The plan focuses on protection of breeding, migratory staging and stopover, and wintering habitats. In Canada, five sites are designated as part of the Western Hemisphere Shorebird Reserve Network, with forty-nine others identified as appropriate for inclusion. The U.S. Shorebird Conservation Plan, developed by the Fish and Wildlife Service, complements the Canadian plan. The plan involves collaboration of governmental agencies and private organizations and is focused on habitat protection. The plan concentrates on fifty species, many with different regional populations, and sets target goals for increasing populations of most species.

The Western Hemisphere Shorebird Reserve Network, launched in 1985, is a conservation strategy focused on key shorebird habitats throughout North and South America. Strategies and operation plans are formulated by the Hemispheric Council consisting of representatives from participating countries and other shorebird conservation groups. The Manomet Center for Conservation Sciences in Manomet, Massachusetts, acts as the implementing organization. The reserve network includes sixty-nine sites in ten countries, with a total area of over 21 million acres. Sites designated as part of the network must meet basic criteria of habitat value and protective management.

### Eurasia, Africa, and the Pacific Region

The Asia–Pacific Migratory Waterbird Conservation Strategy was initiated in 1996 to coordinate national and private efforts to protect migratory waterbirds. This organization has developed action plans for the Central Asian Flyway and East Asian–Australasian Flyway. One critical action plan concerns development of a network of protected wetland sites for migratory shorebirds in the East Asian–Australasian Flyway.

The Agreement on the Conservation of African–Eurasian Migratory

Waterbirds was developed in 1995 in the Hague, the Netherlands, and became effective in 1999 after it was ratified by seven countries from Africa and seven from Eurasia. The agreement sets out guidelines for protection of waterbirds and their habitats, and it applies to 235 species of birds that depend on wetlands, including loons, grebes, pelicans, cormorants, herons, storks, rails, ibises, spoonbills, flamingos, ducks, swans, geese, cranes, waders, gulls, terns, and the Jackass Penguin.

Australia also made international agreements with Japan in 1981, China in 1986, and Korea in 2007 to protect areas used by migratory birds. These treaties list migratory land and waterbird species of particular concern. The United States also has signed treaties of a similar nature with Mexico, Japan, and Russia.

## Control of International Trade

The Convention on International Trade in Endangered Species of Wild Fauna and Flora (CITES) was drafted in 1963 by the IUCN and, after ratification by 80 countries, was implemented in 1975. CITES now has a membership of 172 countries. The convention covers more than 30,000 species of plants and animals. The species covered are listed in three appendices. Appendix I lists species threatened with extinction, with trade permitted only in exceptional cases. Appendix II lists species for which trade must be controlled to avoid exploitation incompatible with survival. Appendix III lists species that are protected in a country or countries that have asked for assistance to control trade. For birds, Appendix I lists 152 species, Appendix II 1268 species, and Appendix III 35 species. Most birds listed are nonmigratory species, but Appendix I lists three shorebirds, seven falcons, eight cranes, and a few other migratory species. Appendix II lists all flamingos, hawks, owls, hummingbirds, and some other birds, many of which are migratory.

## Outlook for Migratory Bird Conservation

Clearly, migratory birds face enormous challenges for survival in a world of rapid climate change. All migratory species depend on the existence of suitable habitats in different geographical areas and at the times appropriate for the different activities of the annual cycle. Some possess a high element of adaptability in their use of these different environments. Many temperate zone short-distance migrants appear to be adjusting ranges and migration schedules to changing climate. Their mobility and responsiveness to changing weather conditions give them the potential for rapid response to climate change.

Many long-distance migrants, especially small songbirds and shorebirds, face enormous challenges from climatic change. Changes in the location and seasonal occurrence of optimal breeding and nonbreeding habitat are often not reflected in the cues through which these species schedule migratory movements. Successful migration also requires that staging and stopover areas be adequate in space and time. To keep the timing of migration synchronized with these environmental conditions requires continual evolutionary fine-tuning. It is evident from contrasting patterns of adjustment of short- and long-distance migrants that some long-distance migrants are slow to adjust to changing conditions in their breeding ranges.

Landscapes changing rapidly in structure and pattern due to climate change combined with human modification are likely to drive some migratory bird populations toward extinction. Avian demography models show that a local population of a species exhibits a critical vulnerability threshold. This threshold represents the landscape state at which the population growth rate becomes sharply negative, leading to a decline toward extinction. At that point it becomes a population sink that can only be maintained by immigration from other areas. For many migratory birds, the seriousness of habitat change may not be apparent until this threshold is reached, at which point population recovery may be impossible.

## Recommendations for Migratory Bird Management

As we have seen earlier in this chapter, major efforts are underway to protect migratory birds. These must be strengthened, and a number of new approaches must be initiated if many of the most remarkable bird migrations are to survive. Some of these are outlined below:

- Expand migratory bird monitoring networks. Our understanding of the timing and volume of bird migration, as well as nonbreeding distributions, is severely deficient in much of eastern and southern Asia, South America, and Africa. Altitudinal and short-distance movements of birds are inadequately recorded throughout the tropics.

- Enhance migratory corridor and stopover networks. With global climate change and increasing human transformation of natural ecosystems, the need for protecting and improving habitats essential to migrating birds is increasing. Especially important are forest habitats in regions of open country, wetlands in continental interiors, and natural areas bordering ocean and desert crossings. In the tropics, where altitudinal movements are extensive, linkages between montane forests and lowland areas used by nonbreeding birds must be maintained. In semiarid regions of the tropics and subtropics, linkages between areas of seasonal use need to be established.
- Employ an anticipatory approach to preserve design and establishment. In areas of changing climate, current preserve locations may become inadequate for breeding and nonbreeding birds. Projections of bird distributions with climate change through the current century, such as those presented in *A Climatic Atlas of European Breeding Birds*, need to be developed for all world regions and used to identify locations of preserve inadequacy.
- Develop a system of assisted range adjustment. The technology of assisting the introduction of breeding stock to habitat areas that are becoming favorable but are isolated geographically needs to be explored in much greater detail. This may be especially important for migratory cliff- and island-nesting seabirds, birds of wetland habitats, birds of fragmented landscapes, and birds of high-mountain areas.
- Increase protection of colonial bird nesting and roosting areas. Eradication programs for introduced predators on colonial seabirds nesting on islands must be expanded. Controls on human harvests of seabirds and their eggs need to be imposed in some regions.

## Summary

About 1.3 percent of bird species have become extinct in recent history, and about 12.0 percent are now endangered. About 8.0 percent of migratory land and freshwater birds are at risk of extinction, but about 44 percent of migratory seabirds are at risk. Migratory species at greatest risk are long-distance temperate–tropical land bird migrants, high-latitude shorebirds, migrants breeding at high elevations and in interior wetlands, nomadic species of semiarid regions, and seabirds. Systems for monitoring populations of breeding land birds now exist in North America and Europe. Several important national and international programs also give protection to migratory birds and their habitats. Many short-distance migrants seem to be responding well to changing climate, but many long-distance migrant songbirds and shorebirds are not showing adequate responses. Improved monitoring of migratory species and expanded networks of preserves covering the full geographical range of these species must be implemented. Techniques for assisting species to colonize new regions with favorable climate also must be developed.

## KEY REFERENCES

BirdLife International. 2008. *Threatened Birds of the World 2008*. BirdLife International, Cambridge, UK.

Doswald, N., S. G. Willis, Y. C. Collingham, D. J. Paim, R. E. Green, and B. Huntley. 2009. "Potential impacts of climatic change on the breeding and non-breeding ranges and migration distance of *Sylvia* warblers." *Journal of Biogeography* 36:1194–1208.

Huntley, B., R. E. Green, Y. C. Collingham, and S. G. Willis. 2007. *A Climatic Atlas of European Breeding Birds*. Durham University, The Royal Society for Preservation of Birds and Lynx Editions, Barcelona.

IUCN (International Union for Conservation of Nature). 2008. *IUCN Red List 2008 for Birds: Climate Change and Continental Drift*. Species Survival Commission Red List Programme, Cambridge, UK.

Kirby, J. S., A. J. Stattersfield, S. H. M. Butchart, M. J. Evans, R. F. A. Grimmett, V. R. Jones, J. O'Sullivan, G. M. Tucker, and I. Newton. 2008. "Key conservation issues for migratory land- and waterbird species on the world's major flyways." *Bird Conservation International* 18:S49–S73.

Mehlman, D. W., S. E. Mabey, D. M. Ewart, C. Duncan, B. Abel, D. Cimprich, R. D. Sutter, and M. Woodrey. 2005. "Conserving stopover sites for forest-dwelling migratory landbirds." *The Auk* 122:1281–1290.

Morrison, R. I. G., Y. Aubrey, R. W. Butler, G. W. Beyersbergen, G. M. Donaldson, C. L. Gratto-Trevor, P. W. Hicklin, V. H. Johnston, and R. K. Ross. 2001. "Declines in North American shorebird populations." *Wader Study Group Bulletin* 94:34–38.

Rich, T. D., C. J. Beardmore, H. Berlanga, P. J. Blancher, M. S. W. Bradstreet, G. S. Butcher, D. W. Demarest, et al. 2004. Partners in Flight: North American landbird conservation plan. Cornell Lab of Ornithology, Ithaca, NY. http://www.partnersinflight.org/.

Sauer, J. R., J. E. Hines, and J. Fallon. 2005. *The North American Breeding Bird Survey, Results and Analysis 1966–2005*. USGS Patuxent Wildlife Research Center, Laurel, MD.

Sekercioglu, C., G. C. Daily, and P. R. Ehrlich. 2004. "Ecosystem consequences of bird declines." *Proceedings of the National Academy of Sciences USA* 101:18042–18047.

U.S. Fish and Wildlife Service. 2004. *A Blueprint for the Future of Migratory Birds*. Migratory bird program strategic plan 2004–2014.

Wormworth, J. and K. Mallon. 2006. Bird species and climate change: The global status report. Version 1.0. Report to World Wide Fund for Nature. http://www.climaterisk.com.au/wp-content/uploads/2006/CR_Report_BirdSpecies ClimateChange.pdf.

# APPENDIX I. COMMON AND SCIENTIFIC NAMES OF SPECIES DISCUSSED IN THE TEXT

## STRUTHIONIFORMES: Struthionidae
Ostrich *Struthio camelus*

## STRUTHIONIFORMES: Casuariidae
Dwarf Cassowary *Casuarius bennetti*

## STRUTHIONIFORMES: Dromaiidae
Emu *Dromaius novaehollandiae*

## SPHENISCIFORMES: Spheniscidae
King Penguin *Aptenodytes patagonicus*
Emperor Penguin *Aptenodytes forsteri*
Adélie Penguin *Pygoscelis adeliae*
Gentoo Penguin *Pygoscelis papua*
Chinstrap Penguin *Pygoscelis antarcticus*
Jackass Penguin *Spheniscus demersus*
Magellanic Penguin *Spheniscus magellanicus*
Humboldt Penguin *Spheniscus humboldti*
Galapagos Penguin *Spheniscus mendiculus*
Macaroni Penguin *Eudyptes chrysolophus*
Rockhopper Penguin *Eudyptes chrysocome*

## GAVIIFORMES: Gaviidae
Red-throated Loon *Gavia stellata*
Common Loon *Gavia immer*

## PODICEPIDIFORMES: Podicipedidae
Western Grebe *Aechmophorus occidentalis*
Australasian Grebe *Tachybaptus novaehollandiae*
Great Crested Grebe *Podiceps cristatus*

## PROCELLARIIFORMES: Diomedeidae
Black-browed Albatross *Thalassarche melanophris*
Sooty Albatross *Phoebetria fusca*

263

## PROCELLARIIFORMES: Procellariidae

Southern Giant-Petrel *Macronectes giganteus*
Northern Giant-Petrel *Macronectes halli*
Northern Fulmar *Fulmarus glacialis*
Southern Fulmar *Fulmarus glacialoides*
Antarctic Petrel *Thalassoica antarctica*
Cape Petrel *Daption capense*
Snow Petrel *Pagodroma nivea*
White-chinned Petrel *Procellaria aequinoctialis*
Sooty Shearwater *Puffinus griseus*
Wedge-tailed Shearwater *Puffinus pacificus*
Balearic Shearwater *Puffinus mauretanicus*

## PROCELLARIIFORMES: Hydrobatidae

Leach's Storm-Petrel *Oceanodroma leucorhoa*
Band-rumped Storm-Petrel *Oceanodroma castro*

## PELECANIFORMES: Sulidae

Peruvian Booby *Sula variegata*
Northern Gannet *Morus bassanus*

## PELECANIFORMES: Pelecanidae

American White Pelican *Pelecanus erythrorhynchos*
Peruvian Pelican *Pelecanus thagus*

## PELECANIFORMES: Phalacrocoracidae

Guanay Cormorant *Phalacrocorax bougainvillii*
Double-crested Cormorant *Phalacrocorax auritus*
European Shag *Phalacrocorax aristotelis*
Antarctic Shag *Phalacrocorax brandsfieldensis*

## CICONIIFORMES: Ardeidae

American Bittern *Botaurus lentiginosus*
Gray Heron *Ardea cinerea*
Great Egret *Ardea alba*
Little Egret *Egretta garzetta*
Cattle Egret *Bubulcus ibis*

## CICONIIFORMES: Ciconiidae

Black Stork *Ciconia nigra*
White Stork *Ciconia ciconia*

## CICONIIFORMES: Threskiornithidae
White Ibis *Eudocimus albus*

## PHOENICOPTERIFORMES: Phoenicopteridae
Greater Flamingo *Phoenicopterus roseus*

## ANSERIFORMES: Anatidae
Pink-footed Goose *Anser brachyrhynchus*
Lesser White-fronted Goose *Anser erythropus*
Emperor Goose *Chen canagica*
Snow Goose *Chen caerulescens*
Brant *Branta bernicla*
Barnacle Goose *Branta leucopsis*
Cackling Goose *Branta hutchinsii*
Canada Goose *Branta canadensis*
Red-breasted Goose *Branta ruficollis*
Tundra Swan *Cygnus columbianus*
Whooper Swan *Cygnus cygnus*
Black Swan *Cygnus atratus*
Mute Swan *Cygnus olor*
Wood Duck *Aix sponsa*
Gadwall *Anas strepera*
American Wigeon *Anas americana*
American Black Duck *Anas rubripes*
Mallard *Anas platyrhynchos*
Mottled Duck *Anas fulvigula*
Blue-winged Teal *Anas discors*
Northern Shoveler *Anas clypeata*
Spot-billed Duck *Anas poecilorhyncha*
Northern Pintail *Anas acuta*
Garganey *Anas querquedula*
Baikal Teal *Anas formosa*
Green-winged Teal *Anas crecca*
Musk Duck *Biziura lobata*
Red-crested Pochard *Netta rufina*
Canvasback *Aythya valisineria*
Redhead *Aythya americana*
Common Pochard *Aythya ferina*
Greater Scaup *Aythya marila*
Lesser Scaup *Aythya affinis*
White-eyed Duck *Aythya australis*
Steller's Eider *Polysticta stelleri*

Spectacled Eider *Somateria fischeri*
Common Eider *Somateria mollissima*
Common Goldeneye *Bucephala clangula*
Common Merganser *Mergus merganser*
Ruddy Duck *Oxyura jamaicensis*

## FALCONIFORMES: Cathartidae
Black Vulture *Coragyps atratus*
Turkey Vulture *Cathartes aura*

## FALCONIFORMES: Pandionidae
Osprey *Pandion haliaetus*

## FALCONIFORMES: Accipitridae
Pacific Baza *Aviceda subcristata*
Mississippi Kite *Ictinia mississippiensis*
Black Kite *Milvus migrans*
Bald Eagle *Haliaeetus leucocephalus*
Steller's Sea-Eagle *Haliaeetus pelagicus*
Lammergeier *Gypaetus barbatus*
Egyptian Vulture *Neophron percnopterus*
Cape Griffon *Gyps coprotheres*
Banded Snake-Eagle *Circaetus cinerascens*
Western Marsh-Harrier *Circus aeruginosus*
Northern Harrier *Circus cyaneus*
Pallid Harrier *Circus macrourus*
Black Harrier *Circus maurus*
Montagu's Harrier *Circus pygargus*
African Marsh Harrier *Circus ranivorous*
Sharp-shinned Hawk *Accipiter striatus*
Cooper's Hawk *Accipiter cooperii*
Black Goshawk *Accipiter melanoleucus*
Levant Sparrowhawk *Accipiter brevipes*
Red-shouldered Hawk *Buteo lineatus*
Broad-winged Hawk *Buteo platypterus*
Swainson's Hawk *Buteo swainsoni*
Red-tailed Hawk *Buteo jamaicensis*
Eurasian Buzzard *Buteo buteo*
Rough-legged Hawk *Buteo lagopus*
Jackal Buzzard *Buteo rufofuscus*
Golden Eagle *Aquila chrysaetos*

## FALCONIFORMES: Sagittaridae
Secretary-bird *Sagittarius serpentarius*

## FALCONIFORMES: Falconidae
Lesser Kestrel *Falco naumanni*
Australian Kestrel *Falco cenchroides*
American Kestrel *Falco sparverius*
Merlin *Falco columbarius*
Eurasian Hobby *Falco subbuteo*
Peregrine Falcon *Falco peregrinus*
Red-footed Falcon *Falco vespertinus*

## GALLIFORMES: Cracidae
Crested Guan *Penelope purpurascens*
Black Guan *Chamaepetes unicolor*

## GALLIFORMES: Phasianidae
Willow Ptarmigan *Lagopus lagopus*
White-tailed Ptarmigan *Lagopus leucura*
Greater Prairie-Chicken *Tympanuchus cupido*

## GRUIFORMES: Gruidae
Blue Crane *Anthropoides paradiseus*
Siberian Crane *Grus leucogeranus*
Sandhill Crane *Grus canadensis*
Common Crane *Grus grus*
Whooping Crane *Grus americana*

## GRUIFORMES: Rallidae
Black Rail *Laterallus jamaicensis*
Virginia Rail *Rallus limicola*
Water Rail *Rallus aquaticus*
Sora *Porzana carolina*
Common Moorhen *Gallinula chloropus*
Eurasian Coot *Fulica atra*

## CHARADRIIFORMES: Haematopodidae
Eurasian Oystercatcher *Haematopus ostralegus*
American Oystercatcher *Haematopus palliatus*

## CHARADRIIFORMES: Recurvirostridae
Black-winged Stilt *Himantopus himantopus*
Black-necked Stilt *Himantopus mexicanus*
American Avocet *Recurvirostra americana*

## CHARADRIIFORMES: Glareolidae
Collared Pratincole *Glareola pratincola*

## CHARADRIIFORMES: Charadriidae
Northern Lapwing *Vanellus vanellus*
Black-bellied Plover *Pluvialis squatarola*
Eurasian Golden-Plover *Pluvialis apricaria*
American Golden-Plover *Pluvialis dominica*
Common Ringed Plover *Charadrius hiaticula*
Piping Plover *Charadrius melodus*
Little Ringed Plover *Charadrius dubius*
Double-banded Plover *Charadrius bicinctus*
Killdeer *Charadrius vociferus*
Mountain Plover *Charadrius montanus*
Eurasian Dotterel *Charadrius morinellus*

## CHARADRIIFORMES: Scolopacidae
Common Sandpiper *Actitis hypoleucos*
Spotted Sandpiper *Actitis macularius*
Green Sandpiper *Tringa ochropus*
Solitary Sandpiper *Tringa solitaria*
Spotted Redshank *Tringa erythropus*
Greater Yellowlegs *Tringa melanoleuca*
Common Greenshank *Tringa nebularia*
Willet *Tringa semipalmata*
Lesser Yellowlegs *Tringa flavipes*
Wood Sandpiper *Tringa glareola*
Common Redshank *Tringa totanus*
Upland Sandpiper *Bartramia longicauda*
Whimbrel *Numenius phaeopus*
Bristle-thighed Curlew *Numenius tahitiensis*
Eskimo Curlew *Numenius borealis*
Eurasian Curlew *Numenius arquata*
Long-billed Curlew *Numenius americanus*
Black-tailed Godwit *Limosa limosa*
Bar-tailed Godwit *Limosa lapponica*
Red Knot *Calidris canutus*
Semipalmated Sandpiper *Calidris pusilla*
Western Sandpiper *Calidris mauri*
Red-necked Stint *Calidris ruficollis*
Temminck's Stint *Calidris temminckii*
Least Sandpiper *Calidris minutilla*
Pectoral Sandpiper *Calidris melanotos*

Dunlin *Calidris alpina*
Curlew Sandpiper *Calidris ferruginea*
Stilt Sandpiper *Calidris himantopus*
Spoon-billed Sandpiper *Eurynorhynchus pygmeus*
Broad-billed Sandpiper *Limicola falcinellus*
Ruff *Philomachus pugnax*
Jack Snipe *Lymnocryptes minimus*
Great Snipe *Gallinago media*
Common Snipe *Gallinago gallinago*
Eurasian Woodcock *Scolopax rusticola*
American Woodcock *Scolopax minor*
Wilson's Phalarope *Phalaropus tricolor*
Red-necked Phalarope *Phalaropus lobatus*
Red Phalarope *Phalaropus fulicarius*

## CHARADRIIFORMES: Laridae
Black-legged Kittiwake *Rissa tridactyla*
Ivory Gull *Pagophila eburnea*
Bonaparte's Gull *Chroicocephalus philadelphia*
Black-headed Gull *Chroicocephalus ridibundus*
Mediterranean Gull *Ichthyaetus melanocephalus*
Ring-billed Gull *Larus delawarensis*
Herring Gull *Larus argentatus*
Lesser Black-backed Gull *Larus fuscus*
Glaucous Gull *Larus hyperboreus*

## CHARADRIIFORMES: Sternidae
Brown Noddy *Anous stolidus*
Sooty Tern *Onychoprion fuscatus*
Little Tern *Sternula albifrons*
Caspian Tern *Hydroprogne caspia*
Arctic Tern *Sterna paradisaea*
Black Skimmer *Rynchops niger*

## CHARADRIIFORMES: Stercorariidae
South Polar Skua *Stercorarius maccormicki*
Brown Skua *Stercorarius antarcticus*
Parasitic Jaeger *Stercorarius parasiticus*

## CHARADRIIFORMES: Alcidae
Common Murre *Uria aalge*
Thick-billed Murre *Uria lomvia*
Razorbill *Alca torda*

Black Guillemot *Cepphus grylle*
Pigeon Guillemot *Cepphus columba*
Kittlitz's Murrelet *Brachyramphus brevirostris*
Marbled Murrelet *Brachyramphus marmoratus*
Ancient Murrelet *Synthliboramphus antiquus*
Cassin's Auklet *Ptychoramphus aleuticus*
Parakeet Auklet *Aethia psittacula*
Least Auklet *Aethia pusilla*
Whiskered Auklet *Aethia pygmaea*
Crested Auklet *Aethia cristatella*
Rhinoceros Auklet *Cerorhinca monocerata*
Atlantic Puffin *Fratercula arctica*
Horned Puffin *Fratercula corniculata*
Tufted Puffin *Fratercula cirrhata*

## COLUMBIFORMES: Columbidae

Passenger Pigeon *Ectopistes migratorius*
Picazuro Pigeon *Patagioenas picazuro*
Common Wood-Pigeon *Columba palumbus*
Lemon Dove *Columba larvata*
Rameron Pigeon *Columba arquatrix*
Delegorgue's Pigeon *Columba delegorguei*
Namaqua Dove *Oena capensis*
Mourning Dove *Zenaida macroura*
Purple-winged Ground-Dove *Claravis godefrida*

## PSITTACIFORMES: Cacatuidae

Slender-billed Black-Cockatoo *Calyptorhynchus latirostris*
White-tailed Black-Cockatoo *Calyptorhynchus baudinii*

## PSITTACIFORMES: Psittacidae

Carolina Parakeet *Conuropsis carolinensis*
Purple-crowned Lorikeet *Glossopsitta porphyrocephala*
Red-capped Parrot *Purpureicephalus spurius*
Mealy Parrot *Amazona farinosa*

## CUCULIFORMES: Cuculidae

Common Cuckoo *Cuculus canorus*
Oriental Cuckoo *Cuculus optatus*
Lesser Cuckoo *Cuculus poliocephalus*
Pallid Cuckoo *Cuculus pallidus*
Fan-tailed Cuckoo *Cacomantis flabelliformis*

Australian Koel *Eudynamys cyanocephalus*
Channel-billed Cuckoo *Scythrops novaehollandiae*
Long-tailed Koel *Eudynamys taitensis*
Yellow-billed Cuckoo *Coccyzus americanus*
Black-billed Cuckoo *Coccyzus erythropthalmus*
Dark-billed Cuckoo *Coccyzus melanocoryphus*
Barred Long-tailed Cuckoo *Cercococcyx montanus*
Shining Bronze-Cuckoo *Chrysococcyx lucidus*
Klass's Cuckoo *Chrysococcyx klaas*

## STRIGIFORMES: Tytonidae
Greater Sooty-Owl *Tyto tenebricosa*
African Grass-Owl *Tyto capensis*

## STRIGIFORMES: Strigidae
Flammulated Owl *Otus flammeolus*
Snowy Owl *Bubo scandiacus*
Ferruginous Pygmy-Owl *Glaucidium brasilianum*
Burrowing Owl *Athene cunicularia*
Tawny Owl *Strix aluco*
Spotted Owl *Strix occidentalis*
Barred Owl *Strix varia*
Short-eared Owl *Asio flammeus*
Boreal Owl *Aegolius funereus*
Elf Owl *Micrathene whitneyi*

## CAPRIMULGIFORMES: Caprimulgidae
Gray Nightjar *Caprimulgus indicus*
Pennant-winged Nightjar *Macrodipteryx vexillarius*

## APODIFORMES: Apodidae
White-throated Needletail *Hirundapus caudacutus*
Common Swift *Apus apus*
Fork-tailed Swift *Apus pacificus*
African Palm-Swift *Cypsiurus parvus*

## APODIFORMES: Trochilidae
Green Violetear *Colibri thalassinus*
Brown Violetear *Colibri delphinae*
Rufous Hummingbird *Selasphorus rufus*
Black-crested Coquette *Lophornis helenae*
White-crested Coquette *Lophornis adorabilis*

Green Thorntail *Discosura conversii*
Fiery-throated Hummingbird *Panterpe insignis*
Snowcap *Microchera albocoronata*
Purple-throated Mountain-gem *Lampornis calolaemus*
Green-crowned Brilliant *Heliodoxa jacula*

## TROGONIFORMES: Trogonidae

Bar-tailed Trogon *Apaloderma vittatum*
Elegant Trogon *Trogon elegans*
Resplendent Quetzal *Pharomachrus mocinno*
Eared Quetzal *Euptilotus neoxenus*

## CORACIIFORMES: Alcedinidae

Sacred Kingfisher *Todiramphus sanctus*
Ruddy Kingfisher *Halcyon coromanda*

## CORACIIFORMES: Meropidae

Swallow-tailed Bee-eater *Merops hirundineus*
White-throated Bee-eater *Merops albicollis*
Rainbow Bee-eater *Merops ornatus*
Southern Carmine Bee-eater *Merops nubicoides*

## CORACIIFORMES: Coraciidae

European Roller *Coracias garrulus*
Dollarbird *Eurystomus orientalis*

## CORACIIFORMES: Bucerotidae

White-thighed Hornbill *Ceratogymna albotibialis*
Black-casqued Hornbill *Ceratogymna atrata*
Wreathed Hornbill *Aceros undulatus*
Knobbed Hornbill *Aceros cassidix*

## PICIFORMES: Lybiidae

Crested Barbet *Trachyphonus vaillantii*

## PICIFORMES: Ramphastidae

Emerald Toucanet *Aulacorhynchus prasinus*

## PICIFORMES: Picidae

Yellow-bellied Sapsucker *Sphyrapicus varius*
Olive Woodpecker *Dendropicos griseocephalus*

## PASSERIFORMES: Eurylamidae
African Broadbill *Smithornis capensis*

## PASSERIFORMES: Pittidae
Noisy Pitta *Pitta versicolor*

## PASSERIFORMES: Cotingidae
Swallow-tailed Cotinga *Phibalura flavirostris*
Bare-necked Umbrellabird *Cephalopterus glabricollis*
Three-wattled Bellbird *Procnias tricarunculatus*

## PASSERIFORMES: Pipridae
Red-capped Manakin *Pipra mentalis*
White-crowned Manakin *Dixiphia pipra*
White-ruffed Manakin *Corapipo altera*

## PASSERIFORMES: Tyrannidae
Sooty Tyrannulet *Serpophaga nigricans*
Ochre-bellied Flycatcher *Mionectes oleagineus*
Tufted Flycatcher *Mitrophanes phaeocercus*
Alder Flycatcher *Empidonax alnorum*
Willow Flycatcher *Empidonax traillii*
Least Flycatcher *Empidonax minimus*
Eastern Phoebe *Sayornis phoebe*
Piratic Flycatcher *Legatus leucophaius*
Scissor-tailed Flycatcher *Tyrannus forficatus*
Black-crowned Tityra *Tityra inquisitor*

## PASSERIFORMES: Alaudidae
Horned Lark *Eremophila alpestris*
Sky Lark *Alauda arvensis*

## PASSERIFORMES: Hirundinidae
Gray-breasted Martin *Progne chalybea*
Tree Swallow *Tachycineta bicolor*
Bank Swallow *Riparia riparia*
Barn Swallow *Hirundo rustica*
Welcome Swallow *Hirundo neoxena*
Cliff Swallow *Petrochelidon pyrrhonota*
Blue Swallow *Hirundo atrocaerulea*

## PASSERIFORMES: Motacillidae

Western Yellow Wagtail *Motacilla flava*
Citrine Wagtail *Motacilla citreola*
White Wagtail *Motacilla alba*
Mountain Wagtail *Motacilla clara*
Richard's Pipit *Anthus richardi*
Mountain Pipit *Anthus hoeschi*
Sprague's Pipit *Anthus spragueii*
Yellow-breasted Pipit *Hemimacronyx chloris*

## PASSERIFORMES: Campephagidae

Ashy Minivet *Pericrocotus divaricatus*
Gray Cuckoo-shrike *Coracina caesia*
Purple-throated Cuckoo-shrike *Campephaga quiscalina*

## PASSERIFORMES: Pycononotidae

Stripe-cheeked Greenbul *Andropadus milanjensis*
Shelley's Greenbul *Andropadus masukuensis*
Eastern Mountain Greenbul *Andropadus nigriceps*

## PASSERIFORMES: Regulidae

Ruby-crowned Kinglet *Regulus calendula*

## PASSERIFORMES: Troglodytidae

Bewick's Wren *Thryomanes bewickii*

## PASSERIFORMES: Mimidae

Gray Catbird *Dumetella carolinensis*
Sage Thrasher *Oreoscoptes montanus*
Brown Thrasher *Toxostoma rufum*

## PASSERIFORMES: Turdidae

Siberian Thrush *Zoothera sibirica*
Scaly Thrush *Zoothera dauma*
Orange Ground-Thrush *Zoothera gurneyi*
Eastern Bluebird *Sialia sialis*
Townsend's Solitaire *Myadestes townsendi*
Veery *Catharus fuscescens*
Gray-cheeked Thrush *Catharus minimus*
Bicknell's Thrush *Catharus bicknelli*
Swainson's Thrush *Catharus ustulatus*
Hermit Thrush *Catharus guttatus*

Black-headed Nightingale-Thrush *Catharus mexicanus*
Wood Thrush *Hylocichla mustelina*
Ring Ouzel *Turdus torquatus*
Eurasian Blackbird *Turdus merula*
Redwing *Turdus iliacus*
Song Thrush *Turdus philomelos*
Brown-headed Thrush *Turdus chrysolaus*
Japanese Thrush *Turdus cardis*
American Robin *Turdus migratorius*
White-chested Alethe *Alethe diademata*

## PASSERIFORMES: Sylviidae

Cameroon Scrub-Warbler *Bradypterus lopezi*
Sakhalin Warbler *Locustella amnicola*
Sedge Warbler *Acrocephalus schoenobaenus*
Marsh Warbler *Acrocephalus palustris*
Great Reed-Warbler *Acrocephalus arundinaceus*
Eurasian Reed-Warbler *Acrocephalus scirpaceus*
Icterine Warbler *Hippolais icterina*
Willow Warbler *Phylloscopus trochilus*
Common Chiffchaff *Phylloscopus collybita*
Wood Warbler *Phylloscopus sibilatrix*
Arctic Warbler *Phylloscopus borealis*
Eastern Crowned Leaf-Warbler *Phylloscopus coronatus*
Western Bonelli's Warbler *Phylloscopus bonelli*
Blackcap *Sylvia atricapilla*
Garden Warbler *Sylvia borin*
Greater Whitethroat *Sylvia communis*
Lesser Whitethroat *Sylvia curruca*
Sardinian Warbler *Sylvia melanocephala*
Little Grassbird *Megalurus gramineus*

## PASSERIFORMES: Polioptilidae

Blue-gray Gnatcatcher *Polioptila caerulea*

## PASSERIFORMES: Muscicapidae

Spotted Flycatcher *Muscicapa striata*
European Pied Flycatcher *Ficedula hypoleuca*
Collared Flycatcher *Ficedula albicollis*
White-starred Robin *Pogonocichla stellata*
Sharpe's Akalat *Sheppardia sharpei*
Common Nightingale *Luscinia megarhynchos*
Thrush Nightingale *Luscinia luscinia*

Bluethroat *Luscinia svecica*
Siberian Blue Robin *Luscinia cyane*
Black Redstart *Phoenicurus ochruros*
Common Redstart *Phoenicurus phoenicurus*
Little Forktail *Enicurus scouleri*
Black Scrub-Robin *Cercotrichas podobe*
Whinchat *Saxicola rubetra*
African Stonechat *Saxicola torquatus*
Northern Wheatear *Oenanthe oenanthe*

## PASSERIFORMES: Rhipiduridae
Gray Fantail *Rhipidura albiscapa*
Rufous Fantail *Rhipidura rufifrons*

## PASSERIFORMES: Monarchidae
White-tailed Crested-Flycatcher *Elminia albonotata*
African Paradise-Flycatcher *Terpsiphone viridis*
Japanese Paradise-Flycatcher *Terpsiphone atrocaudata*
Black-faced Monarch *Monarcha melanopsis*
Leaden Flycatcher *Myiagra rubecula*
Restless Flycatcher *Myiagra inquieta*

## PASSERIFORMES: Petroicidae
Flame Robin *Petroica phoenicea*

## PASSERIFORMES: Pachycephalidae
Olive Whistler *Pachycephala olivacea*
Rufous Whistler *Pachycephala rufiventris*

## PASSERIFORMES: Timaliidae
African Hill Babbler *Pseudoalcippe abyssinica*
Spot-throat *Modulatrix stictigula*

## PASSERIFORMES: Acanthizidae
White-throated Gerygone *Gerygone olivacea*

## PASSERIFORMES: Paridae
Mountain Chickadee *Poecile gambeli*
Great Tit *Parus major*
Eurasian Blue Tit *Cyanistes caeruleus*

## PASSERIFORMES: Sittidae
Red-breasted Nuthatch *Sitta canadensis*

## PASSERIFORMES: Certhiidae
Brown Creeper *Certhia Americana*

## PASSERIFORMES: Nectariniidae
Banded Sunbird *Anthreptes rubritorques*
Eastern Olive Sunbird *Cyanomitra olivacea*

## PASSERIFORMES: Pardalotidae
Striated Pardalote *Pardalotus striatus*

## PASSERIFORMES: Meliphagidae
Yellow-faced Honeyeater *Lichenostomus chrysops*
Crescent Honeyeater *Phylidonyris pyrrhopterus*
Red Wattlebird *Anthochaera carunculata*

## PASSERIFORMES: Oriolidae
Eurasian Golden Oriole *Oriolus oriolus*

## PASSERIFORMES: Laniidae
Tiger Shrike *Lanius tigrinus*
Brown Shrike *Lanius cristatus*
Loggerhead Shrike *Lanius ludovicianus*

## PASSERIFORMES: Malaconotidae
Black-fronted Bushshrike *Telephorus nigrifrons*
Fuelleborn's Boubou *Laniarius fuelleborni*

## PASSERIFORMES: Ptilonorhynchidae
Golden Bowerbird *Prionodura newtoniana*

## PASSERIFORMES: Corvidae
Steller's Jay *Cyanocitta stelleri*
Blue Jay *Cyanocitta cristata*
Rook *Corvus frugilegus*

## PASSERIFORMES: Sturnidae
Chestnut-cheeked Starling *Sturnia philippensis*
European Starling *Sturnus vulgaris*

White-cheeked Starling *Sturnus cineraceus*
Waller's Starling *Onychognathus walleri*
Kenrick's Starling *Poeoptera kenricki*

## PASSERIFORMES: Estrildidae
Red-faced Crimson-wing *Cryptospiza reichenovii*

## PASSERIFORMES: Vireonidae
Bell's Vireo *Vireo bellii*
Black-capped Vireo *Vireo atricapilla*
Yellow-green Vireo *Vireo flavoviridis*
Cassin's Vireo *Vireo cassini*

## PASSERIFORMES: Fringillidae
Chaffinch *Fringilla coelebs*
Purple Finch *Carpodacus purpureus*
House Finch *Carpodacus mexicanus*
White-winged Crossbill *Loxia leucoptera*
European Greenfinch *Carduelis chloris*
Pine Siskin *Spinus pinus*
European Serin *Serinus serinus*
Black-throated Canary *Serinus atrogularis*
Cape Siskin *Pseudochloroptila totta*
Drakensberg Siskin *Pseudochloroptila symonsi*

## PASSERIFORMES: Drepanididae
Iiwi *Vestiaria coccinea*
Apapane *Himatione sanguinea*

## PASSERIFORMES: Parulidae
Blue-winged Warbler *Vermivora pinus*
Golden-winged Warbler *Vermivora chrysoptera*
Tennessee Warbler *Vermivora peregrina*
Orange-crowned Warbler *Vermivora celata*
Nashville Warbler *Vermivora ruficapilla*
Northern Parula *Parula americana*
Yellow Warbler *Dendroica petechia*
Chestnut-sided Warbler *Dendroica pensylvanica*
Magnolia Warbler *Dendroica magnolia*
Cape May Warbler *Dendroica tigrina*
Black-throated Blue Warbler *Dendroica caerulescens*
Yellow-rumped Warbler *Dendroica coronata*
Golden-cheeked Warbler *Dendroica chrysoparia*

Black-throated Green Warbler *Dendroica virens*
Blackburnian Warbler *Dendroica fusca*
Yellow-throated Warbler *Dendroica dominica*
Kirtland's Warbler *Dendroica kirtlandii*
Prairie Warbler *Dendroica discolor*
Palm Warbler *Dendroica palmarum*
Bay-breasted Warbler *Dendroica castanea*
Blackpoll Warbler *Dendroica striata*
Cerulean Warbler *Dendroica cerulea*
Black-and-white Warbler *Mniotilta varia*
American Redstart *Setophaga ruticilla*
Prothonotary Warbler *Protonotaria citrea*
Worm-eating Warbler *Helmitheros vermivorum*
Swainson's Warbler *Limnothlypis swainsonii*
Ovenbird *Seiurus aurocapilla*
Northern Waterthrush *Seiurus noveboracensis*
Louisiana Waterthrush *Seiurus motacilla*
Kentucky Warbler *Oporornis formosus*
Connecticut Warbler *Oporornis agilis*
Mourning Warbler *Oporornis philadelphia*
MacGillivray's Warbler *Oporornis tolmiei*
Common Yellowthroat *Geothlypis trichas*
Hooded Warbler *Wilsonia citrina*
Wilson's Warbler *Wilsonia pusilla*
Canada Warbler *Wilsonia canadensis*
Slate-throated Redstart *Myioborus miniatus*
Collared Redstart *Myioborus torquatus*

## PASSERIFORMES: Thraupidae
Summer Tanager *Piranga rubra*
Western Tanager *Piranga ludoviciana*
Common Bush-Tanager *Chlorospingus ophthalmicus*
Fawn-breasted Tanager *Pipraeidea melanonota*

## PASSERIFORMES: Emberizidae
Yellow-thighed Finch *Pselliophorus tibialis*
Spotted Towhee *Pipilo maculatus*
Eastern Towhee *Pipilo erythrophthalmus*
Cassin's Sparrow *Aimophila cassinii*
Chipping Sparrow *Spizella passerina*
Clay-colored Sparrow *Spizella pallida*
Brewer's Sparrow *Spizella breweri*
Field Sparrow *Spizella pusilla*

Vesper Sparrow *Pooecetes gramineus*
Lark Sparrow *Chondestes grammacus*
Lark Bunting *Calamospiza melanocorys*
Savannah Sparrow *Passerculus sandwichensis*
Grasshopper Sparrow *Ammodramus savannarum*
Baird's Sparrow *Ammodramus bairdii*
Henslow's Sparrow *Ammodramus henslowii*
Le Conte's Sparrow *Ammodramus leconteii*
Nelson's Sharp-tailed Sparrow *Ammodramus nelsoni*
Fox Sparrow *Passerella iliaca*
Song Sparrow *Melospiza melodia*
Lincoln's Sparrow *Melospiza lincolnii*
Swamp Sparrow *Melospiza georgiana*
White-throated Sparrow *Zonotrichia albicollis*
White-crowned Sparrow *Zonotrichia leucophrys*
Dark-eyed Junco *Junco hyemalis*
McCown's Longspur *Calcarius mccownii*
Chestnut-collared Longspur *Calcarius ornatus*
Chestnut-eared Bunting *Emberiza fucata*
Yellow-breasted Bunting *Emberiza aureola*
Reed Bunting *Emberiza schoeniclus*
Corn Bunting *Emberiza calandra*
Lark-like Bunting *Emberiza impetuani*
Snow Bunting *Plectrophenax nivalis*

## PASSERIFORMES: Cardinalidae
Rose-breasted Grosbeak *Pheucticus ludovicianus*
Glaucous-blue Grosbeak *Cyanoloxia glaucocaerulea*
Lazuli Bunting *Passerina amoena*
Painted Bunting *Passerina ciris*
Dickcissel *Spiza americana*

## PASSERIFORMES: Icteridae
Bobolink *Dolichonyx oryzivorus*
Red-winged Blackbird *Agelaius phoeniceus*
Eastern Meadowlark *Sturnella magna*
Western Meadowlark *Sturnella neglecta*
Common Grackle *Quiscalus quiscula*
Great-tailed Grackle *Quiscalus mexicanus*
Brown-headed Cowbird *Molothrus ater*
Baltimore Oriole *Icterus galbula*

*Source:* Clements, J. F. 2007. *The Clements Checklist of Birds of the World*, 6th ed. Comstock Publishing Associates (Cornell University Press), Ithaca, NY.

# INDEX

Page numbers followed by "f" and "t" indicate figures and tables.